FAUNA OF KAKADU AND THE TOP END

Denise Goodfellow came to the Top End, in 1975 as a classic guitar teacher, then shot buffalo for two years. Elected to Darwin City Council, in 1981 in an attempt to save local mangroves, she tried to represent Aboriginal people, recognising that they had more urgent priorities than potholes. She gained the women's confidence by running beauty courses on the local reserve, and then attempted to tackle problems such as non-attendance by police at fights, rapes and burglaries.

In the company of Aboriginal women, Ms Goodfellow spent four hours wading in a billabong inhabited by crocodiles, in a successful attempt to catch water python for the reserve's approaching open day, at which bush tucker would be served. Her feat was reported widely in local papers bringing about a spate of demands for her prosecution. The President of the reserve adopted Ms Goodfellow in an attempt to protect her.

In 1985 Denise co-produced, with Hilary Thompson, *Common Birds of the Darwin Area.* Since then she has written articles on monsoon forest, birds, spiders, and cross-cultural relationships, and has delivered papers on sustainable tourism at conferences, mostly with instructive titles like *Male Black Whip Snakes are very well-equipped.* Her book, *Birds of Kakadu and the Top End,* will be published early next year.

Ms Goodfellow works as a biological and ecotourism consultant, and as a specialist tour guide. In the former role she carries out surveys of flora and fauna, often alone, in the bush. She has shared her sleeping bag with nasty spiders, encountered large crocodiles and venomous snakes, and chased buffalo and amorous miners out of her campsite. She guides mainly for individual naturalists, but also for organizations such as Society Expeditions and the Academy of Natural Sciences, Philadelphia. Her chequered academic background has ranged from environmental science to post-graduate work on the sociology of tourism.

Denise Goodfellow
PO Box 39373
WINNELLIE NT 0821

and
Wakefield Press
PO Box 2266
Kent Town
South Australia 5071

First edition October, 1993.

Cover design by Denise Goodfellow
Printed by Colemans Printing, 86 Cavenagh Street, Darwin 0800.

ISBN 1 86254 306 2

FAUNA of KAKADU and the TOP END

Denise Goodfellow

Denise Goodfellow
in association with Wakefield Press

FOREWORDS

It has been suggested that Charles Darwin had a supreme power of seeing and thinking what the rest of the world had overlooked. Darwin's own characterization in his autobiographical notes called LIFE, state "I think I am superior to the common run of men in noticing things which easily escape attention, and in observing them carefully."

Careful observations, such as those made by Darwin, take the rest of us out of the shadows and into real life. And so it is with Denise Goodfellow's ***Fauna of Kakadu and the Top End.*** The animals are brought out of the shadows, becoming alive with patterns, size and habits, tied in with the environment and Aboriginal people.

On our first day in Charles Darwin's namesake city, Denise insisted we borrow her car. It turned out to be a cube-shaped Willys jeep clone, with a Peugot engine, the whole lot assembled in India and named MAHINDRA. A sort of field biologist's version of a "world" car, we didn't see another vehicle resembling it during the remainder of our time in the Top End. This unexpected availability of motorized personal transport opened the door to a brief foray into some of the habitats close to Darwin. The beaches were deserted except for myriad bird species, many of them new to us. We finally stopped at East Point for a look at a rocky coastline and a dense population of Agile Wallabies. The next day we went bush with Denise. Eventually we added over a hundred species to our bird list. We also learned that the abundance and diversity of fauna of the Top End, is truly remarkable.

Structurally, the north end of the Northern Territory consists of hundreds of miles of coast interrupted by numerous mangrove-dominated estuaries. Inland are monsoon forests, wet forests, dry forests, savannah woodlands, and rocky escarpments. There are also vast wetlands creating a green sea stretching to the horizon, an awsome sight when viewed from rocky heights, such as Ubirr Rock overlooking the East Alligator River in Kakadu National Park. They have, for good reason, been designated as wetlands of international significance by the RAMSAR convention. Equally exciting to the wetland ecologist are the paperbark wetlands with big water lilies *(Nymphea violacea)* and little water lilies *(Nymphoides* species) dotting the understorey waters. Forested wetlands are declining worldwide and require long time periods to be restored. To see such splendid intact examples is truly a delight.

Splice all these systems together with a network of undammed rivers and streams which swell over lowlands in the rainy season and recede to flow sedately within defined channels during dry seasons, leaving behind the billabongs, and you have some of the features which provide the bricks and mortar for the fauna of Kakadu and the Top End. Now introduce a fauna which includes hundreds of species of birds and plants, all three major groups of mammals (egg-laying, marsupial and placental) and a magnificent array of amphibians and reptiles, and you have the Top End. Of course the current animal census can only include known species. On any given sampling foray a previously unknown species might well turn up. In the Top End, the opportunity to see the known is of enormous interest and the experience is magnified by the potential to encounter the unknown.

Fauna of Kakadu and the Top End provides a stepping stone toward experiencing the best of the Northern Territory. Descriptions are straightforward and coherent, and uncluttered with excessive technical terms. Yet the book is comprehensive enough to appeal to visiting scientists, serious amateurs and all the naturalists endemic to the Top End.

Morphological characteristics of animal species is just the beginning. Add to this, descriptions of skeletons and scats (animal poo as the author refers to it in the name of clarity for non-scientists), Aboriginal information, plus habitat and sex life explanations for the various groups. Infused into the science and natural history is practical advice on fauna watching, important suggestions for respecting the rights and privacy of wildlife, and most importantly, the wit and humor of Ms. Goodfellow, who is as unique as her car. Incidentally, I have never been in the field with a better naturalist, anywhere.

How best to rub up against the Top End and its wildlife? Throw together the camping gear and head for the areas described in this book. A 4 wheel-drive helps and it's even better to be accompanied by an experienced Top End naturalist and/or this book. Investigate all movements in the bush and stop for all road kills. You never know when you might find an interesting and hard-to-see species of snake inside a dead King Brown! Be aware that it will probably rain outside of the dry season, and that the mozzies (and midges) may be out in droves. This means you may want to stay in your tent at night. However to endure these minor discomforts is a small price for experiencing Top End wildlife delights. Just be sure the cook leaves the Vegemite at home!

Darwin is fortunate to have a naturalist of Ms. Goodfellow's abilities to chronicle Top End wildlife. We can only hope that future works include spiders, insects, plants etc.

*Fred C. Weinmann, Ph. D**
Wetland Ecologist
Issaquah, Washington USA

*Dr. Weinmann has been with the United States Environmental Protection Agency for twenty years and is currently working on the forest protection report for President Clinton.

Fauna of Kakadu and the Top End is an unusual book. Interesting and chatty, it is also an accurate reference reference book to be used when more scientific differentiation between species is needed. Denise Goodfellow is known locally as being thoroughly conversant with what lives where in the Top End of the Northern Territory and she has made a point of bringing her love and knowledge of our wildlife to us locals and tourists alike.

Denise Goodfellow herself is one of the best wildlife reference books I know and she is always ready to share her knowledge, in a down to earth way. She is, in addition, an accomplished artist. In her exact portrayals she again demonstrates her love and care of wildlife.

We are very lucky in the Top End in that we still have plenty of small wildlife around, and the most exciting part is that you never know what you'll find on the back verandah one morning, in the carport, or next to the chookyard. Now we have ***Fauna of Kakadu and the Top End*** to tell us accurately what bush creature has paid us a visit.

Noel Padgham-Purich, Member of the Legislative Assembly, Northern Territory.

Denise Goodfellow's no-nonsense approach to life, people and causes has raised a few hackles over the years, but has earned for her love and respect from those of us who truly know her.

Denise was one of the first persons to realise, and to understand the value, in both human and dollar terms, of eco-tourism, which she pioneered in the Darwin region. Her contributions to it are recognised nationally and internationally. She has now produced what I would describe as an 'Eco-tourists's Handbook'. I commend the author for her initiative and effort and recommend the book to all lovers of nature.

George Brown, Lord Mayor of Darwin.

PREFACE

A friend,when she heard I was writing a nature guide, asked me to ensure that it would be comprehensive enough to enable her to identify a Frilled Lizard from a fragment of bald tyre at fifty paces in bad light. Unfortunately no book could compensate for her bad eyesight!

However *Fauna of Kakadu and the Top End* should enable most people to identify over 120 species of mammal, reptile and frog found in the Top End of the Northern Territory, Australia, and elsewhere. I have chosen the species herein because they are common, or interesting, or because people have asked me about them. Because people have usually wanted more than just an identification, I have also included other information, such as local Aboriginal names, ecology, taxonomy, palaeontology, and lots of anecdotes.

I hope that my book will encourage both locals and visitors, to get out and look around them. I also hope it will influence those in the wider tourism industry who believe that knowledge is of little importance. The extent of this attitude was brought home to me by a number of experiences. For instance once a man who had spent three hours on my bush tucker course, and had no other knowledge of flora, won a job teaching trainee tour-guides about these subjects. On another occasion a guide leading two Italian photo-journalists from the *National Geographic* told me he didn't know why he kept getting such work, as he didn't know the first thing about wildlife!

Yet nature-oriented visitors desire accurate information, and there is much evidence to show that visitors are disappointed, if not antagonised by such ignorance. Between 60 and 80% of nature-oriented visitors have a tertiary education, and I have found that even on ordinary bus tours there are often doctors, anthropologists and botanists. I hope my book will indicate to tour operators the depth of knowledge that many of these visitors desire.

Some have queried the inclusion of sex in this book. In fact one media story made the book out to be a sex manual! To put the record straight this information is included for a couple of very good reasons. Firstly virgin birth is rare among vertebrates. Put bluntly, no sex usually means no babies. Secondly, the people I guide acknowledge that all animals, unlike Bambi and Co., have sex organs of some kind. They do not see the provision of such information as unusual or suspect.

Who are these perverts, that accompany me on my forays into the bush? Eighty percent are over 45 years old, most are women, almost all are well-educated and well-travelled, and many have raised children, survived serious illnesses, divorce and the death of loved ones.

Their experiences have not only made them open-minded and accepting, but kind, concerned, and honest as well. Yet in Western society people with these qualities are often considered 'suckers', although this attitude is slowly changing, as is shown by the establishment of codes of ethics in business. However tourism with its rigid infrastructure, has a long way to go before it meets the desires and fits the values of these people.

I dedicate ***Fauna of Kakadu and the Top End*** to those older women, and men, of questioning minds, who strive to learn, understand and to communicate. I include here the wonderful Aboriginal woman who adopted me as her sister-in-law, and who I cannot name because, to my everlasting sorrow, she is now dead.

Denise Goodfellow
September, 1993

CONTENTS

About this book

This book is about terrestrial vertebrate fauna of the Top End, the northern third of the Northern Territory. It will also help the visitor elsewhere in northern Australia. I have not tried to cover all animals. Specialised publications are listed in the bibliography for those who want them.

The large format of this book has allowed me to use a bigger type-size and pictures than are usually found in nature-guides, thus making the book more user-friendly to those who have less than perfect eyesight. The soft cover means the book can be rolled in order to fit into a backpack. I hope the stitched binding will enable it to withstand some rough handling.

Maps are on the inside of the front and back covers. They cover all areas mentioned in the book. The section on parks gives brief detail on roads, habitats and distances from major towns.

The sections on how to watch animals and identification of skeletons, tracks and animal poo will introduce readers to non-intrusive ways of learning about animals.

There is a section on each major group which includes information on palaeontology, taxonomy, breeding behaviour etc. This is followed by entries of related species which include descriptions, habits, habitats and areas where the animal is likely to be found, its status, whether it is easily seen or otherwise, and if it is solitary or found in groups. I have used those names in common usage. However many lack such names in which case only scientific and Aboriginal names are supplied.

With most mammals I have attempted to relate size to an animal that the reader may be more familiar with, eg mouse, rat, cat or dog. Measurements are for adults and immatures are often much smaller. Many entries have a section titled Similar and Related Species, which covers species that may be confused with the main entry. The Notes section deals with other information such Aboriginal names.

I have attempted to do away with heavy scientific jargon. Some will criticise me for this, others will say that there is still too much. You cannot please everyone.

Aboriginal information

Aboriginal people have strong ties to the land. Yet their names for animals, and their information, are not usually included in natural history books, an ommission I find odd. Consequently I have attempted to include as much as possible. Aboriginal names included in my book, are mainly Gunwinngu, from Western Arnhemland and I hope to expand this information in further editions.

Pronounciation of some Aboriginal vo a as in h*u*t ollows. ee as in th*eir*
dj as in *de*w
tj as in *tu*ne
ng as in si*ng*
ny as in o*ni*on

u as in p*u*t
aa as in sm*ar*t
ayi as in p*ie*

ey as in m*ay*

Aboriginal people in southern Australia call themselves Koori. In Eastern Arnhemland they call themselves Yolgnu. Western Arnhemland people, such as the Gunwinggu, call themselves Binintj. Rather than use other names inappropriately I have used the term 'indigenous' or 'Aboriginal' when referring collectively to Top End Aboriginal people. Otherwise I have referred to them by the name of their language group, ie Gunwinggu.

The Gunwinggu give specific names to conspicuous animals and to those which are of religious significance, important food items or useful in other ways. Some of these animals have many different names although species similar in appearance may all be lumped under one name. For instance most rats are called Mulbu, as are those carnivorous maruspials that look like rats. In keeping with the tone of the book I have included information from my adopted Aboriginal relatives on catching and cooking animals.

A STATEMENT ON CONSERVATION

Indigenous people lived in relative harmony with their environment for thousands of years. They had little choice because the environment imposed limits.

For instance, water python were hunted by wading in billabongs. There was a limit to the number of water python one could catchif there was a lot of water. Then there were crocodiles. The longer one was in the water the greater the danger. Many billabongs were ephemeral and dried up during the late dry season, a bottleneck that limited the number of water pythons anyway.

Utilisation of resources was also controlled by tradition. Some people had water python dreaming for example, and because of this relationship they were bound to protect this snake and its habitat. This system where some could kill and others had to protect a species, applied to a whole range of animals.The welfare of the individual snake did not matter. However the species and its habitat were of paramount importance.

In today's society we are not limited by our immediate environment. We farm intensively or truck our food in and we build dams to supply water to large populations. We can stand the heat or cold because we live and work in houses with fans, heating and air-conditioning. We attempt to fit the environment to us. If we had to rely only on those natural resources in our own little part of the world we would learn, as Aboriginal people did, to fit our population and consumption to the local environment. We too, would have to conserve through lack of choice. And that day is coming.

Conservation, as Western society practises it, is vastly different from that practised by Aboriginal society. It separates human beings and the natural environment by attempting to make the latter sacrosanct and therefore untouchable. We are told that it is wrong to eat kangaroo (after all it is Australia's national symbol) and that we shouldn't keep native animals as pets, that we shouldn't pick native flowers. Instead we eat beef and lamb, keep cats, and grow roses. All have environmental impacts but most are removed from our sight.

These attitudes affect indigenous people, coming between them and their relationship to the natural environment. I remember a letter to the *Northern Territory News* in October, 1983, stating that 'Aboriginal people do not know any better than to eat protected wildlife'. Aboriginal people were embarrassed by this attitude and gradually at social functions they ceased to serve 'bush tucker'. Often their children see such food as uncivilised, at the same time that Western society is making it a marketable commodity.

Adults, as my adopted Aboriginal son and his wife have commented sadly to me, are then unable to pass traditional skills to them. A society rarely dies with blood running in the streets. Rather it chokes in embarrassed silence as children turn from the ways of their parents.

Yet, recently, I've seen indigenous people go back to their elders and ask for information they once rejected, but now value.

We could possibly turn back to the old ways as well, to the more sustainable lifestyles of our parents and grandparents. Notwithstanding all the clearing and the lack of pollution controls on industry, they had a far less devastating effect on the environment than we have now. For example, did you know that in 1947 the average home had about four people living in about four rooms. Now the average home has about six rooms, but only three people living there. So sewerage, roads, power supply and public and private transport need to travel much further to serve the same number of people.

Yet as the need to conserve grows, our ability to make real changes to our lifestyles shrinks, as we become part of massive web of informal and formal bureaucracies formed by associations, business and government. A rigid infrastructure develops which creates career paths for some. but over time this infrastructure becomes inward-looking, reducing the legitimacy, flexibility and innovation needed to keep it alive. The conservation movementhas become part of this process.

It is easy to become disillusioned, but rather than throw up your hands in despair, do something about conservation in a way that gives you some control.

There are many ways of doing this. Firstly, I suggest that you get rid of your cat. Now of course this has ramifications, as does all conservation. After all you can't carry the flag without carrying the can as well! So, everyone in the house thinks you are horrible, if not weird. Then, one morning you find a Blue-tongued Lizard eating the dog food. You call your children, and together you wonder over this wild animal, so intent on its food it ignores you all.

You may read what you like into the lizard's lack of fear. You may even find yourself going a little sooky and thinking, 'It trusts me!' The next day you suddenly find yourself hosing the neighbour's cat to keep it out of your yard and away from Your Lizard. And your children are writing hate letters to the local newspaper about cats. You may have been recycling your beer cans for years. However I bet you probably never saw recycling as a 'life or death' matter. Yet you really care whether this lizard lives or dies, especially at the claws of a cat.

You can extend this process, by tackling developers who cut down old trees. So far they have not been greatly constrained by by-laws, or emotive arguments, so look at the economic ramifications. It could be costly if a regular user of their area develops skin cancer and decides to sue. Some Sydney councils are taking this threat seriously and so should government and business in the Top End.

BEFORE YOU START OUT

Regulations

A permit is required to collect or keep vertebrates in the Northern Territory, although there are exceptions. Ring the Conservation Commission if you are interested. Interference with dead animals is illegal (in fact I was told I could be prosecuted for showing people dead snakes, even if I didn't remove them from the area). Take note, overseas visitors, whose laws allow you to do things for which we would be flayed alive and nailed to the nearest signpost.

Unfortunately, as far as I can discover, nobody has been prosecuted for the common practice of ramming crocodiles to make them jump. Tour operators do it to 'give visitors a thrill'. Incidentally, at least one of these operators advertises that his tours are 'environmentally-sensitive'. Other unsavoury practices carried out by some tour operators, and others include stoning crocodiles or poking them with sticks, scaring flying-fox colonies to make them fly, intruding on the nesting colonies of waterbirds, and stampeding wallabies. These actions are all illegal.

Concerned people should contact the Northern Territory Conservation Commission, Australian Nature Conservation Agency, Darwin City Council, Northern Territory Tourist Commission and Consumer Affairs. Their numbers are given at the back of this book. Follow up with letters to newspapers and relevant ministers. You may be the only ones capable of bringing about action.

Climate and wildlife

The Top End has a monsoonal climate. Non-Aboriginal Top Enders usually divide it into a wet season (December-March) and a dry season May -September). The months between the end of the dry and the beginning of the wet season (October-December) are known as the buildup. The months between the end of the wet and the beginning of the dry (March - April) feature winds called the knock-em-downs. Some years seasons are shorter or longer. I remember one wet season which began in August, unfortunately for a colleague and I who were camped on a dried up dam. Next April we went out bush thinking the end of the wet was in sight. Down it came again. Unfortunately we had been dropped into a remote area by helicopter and couldn't get out. But that's another story.

Gunwinguu and related Aboriginal people have a different system in which they recognise six seasons. They call the wet season Gudjewg. It features monsoonal downpours and occasional cyclones and floods. Rainfall averages 1500 mm.

Us locals love the wet. It is cooler than the months on either side, and the country is emerald-green - a great time for photographers. There is often heavy rain but it is warm. Wear a broad-brimmed hat (to keep water out of your eyes) and gumboots, especially when going into monsoon forest where leeches may be prevalent.

Although most animals are dispersed during the wet season, frogs can often be seen hopping on the roads at night, and snakes, some native mammals and nocturnal birds come out to feast on them. Frilled Lizards can be seen near or on the road during the day.

The wet is followed by another period of high humidity, high temperatures (up to 36°C), but little rain. During this time we have violent storms and gusts of wind, the 'knock-em-downs'. Many birds are breeding and frogs and their predators are still evident. Wildflowers are a prominent and lovely feature of the landscape at this time, which is called Banggereng by Gunwinggu people.

The dry season is the coolest and driest part of the year. There are snakes, and marsupials such as Brush-tailed Phascogale and kangaroos, on the roads at night, . However the most common animal seems to be 'the tourist'. Gunwinggu people divide this time of year into two parts. The first part (May-July) is Yegge. The second part which is colder and drier is Wurrgeng.

Bushfires are prevalent during the dry, and the sky is covered by smoke-haze which has the amazing effect of making the setting sun apear like a giant pink lollipop! However locals often find the dry 'boring'. Blue skies every day. Ho hum.

To many long-term Top End residents the build-up, that period following the dry season, is the best time of the year. Purple-black cumulo-nimbus, shot through with bolts of lightning, tower into the jetstream backed by orange-red sunsets and cobalt skies. It is hot and humid, a drawback for most visitors. Still October-December is nature-time, with many animals visible. This is because there is so little water animals become concentrated around waterholes. Also many are looking for mates and, like most males, they like to make themselves obvious. When the rains come, they cause cracking clays on the floodplains to swell, forcing their inhabitants out into the open.

This time of year is again divided into two by Gunwinggu people. The first part which is hot and dry, is called Gurrung. The second part, the time of storms, is Gunumeleng.

Habitats

The Top End contains several different habitats. The most common are listed below.

Rocky outcrops
A good example of these are the collections of boulders seen along the Arnhem Highway near the Mary River. They contain a mixture of plants, some belonging to open forest and others of monsoon forest origin. These plants and the rocks provide habitat for several species of rock-dwelling lizards and some mammals such as Echidna and Northern Quoll.

Escarpment and ranges
Apart from the Arnhemland escarpment there are several ranges of rocky hills. They provide a range of habitats from pockets of monsoon forest, gallery forest along streams, heathland, and savanna woodland. There are several species of animal confined to the escarpment country eg the Giant Cave Gecko and Black Wallaroo and some species of frog.

Savanna woodland
This woodland is low in height with trees spread far apart, and a grassy understorey. Trees are mainly eucalypt. Savanna woodland is found on hills and on plains. Different species of animal may be found in each eg Kakadu Dunnart in the hills and Red-cheeked Dunnart on the plains. Antilopine Wallaroo are common in these habitats.

Open forest
This habitat is dominated by mid-high to tall eucalypts, often two species, Darwin Stringybark *(Eucalyptus tetrodonta)* and Woollybutt *(Eucalyptus miniata)*. The understorey may be grassy or herbaceous. Sugar gliders and other marsupials are found in these areas, along with an assortment of reptiles and frogs.

Closed/monsoon/gallery forest
This type of forestcan be complex with many different species of plant and a dense understorey, or simple with just a few species, eg bamboo, and a clear understorey. It is often known as vine-forest and can be wet, as at Litchfield Park and Holmes Jungle, or dry as around the Daly River. Gallery forest is often dominated by tall paperbarks and contains many of the species of animal found in open forest. Agile Wallabies are often found in monsoon forest with a clear understorey.

Floodplains/wetlands
These vast areas are inhabited by a dozens of species of frogs and file snakes, and turtles. The animals inhabiting these areas change with the season. For instance Common Planigales and Dusky Rats are found during the dry season but are largely flooded out during the wet season.

Watching animals

Many animals can be identified by careful viewing. Experiment by walking or driving quietly with frequent stops (cars make good hides). Keep an eye on a visible bird or mammal eg kangaroos, or birds on wires overhead and stopmoving when they show visible signs of alarm. (Bird-spotting is covered in my next book, *Birds of Kakadu and the Top End*).

All your stalking may be undone if you wear bright-coloured clothing, perfume, or talk too much. Animals (except for most birds) can easily detect the scent of humans, and especially mosquito repellant and perfume. The very serious animal observer may try a rubdown with the new leaves of paperbarks or even mud. It makes a wonderful talking-point even if you don't see anything.

You can see much wildlife along the highways of the Top End, including snakes, goannas, carnivorous marsupials such as Northern Quoll, kangaroos, and introduced animals such as horses and water-buffalo. Spotlighting along the roads at night is a good way to see animals. However avoid shining powerful lights in their eyes if possible. You wouldn't like it and neither do they. A dim torch held to one side is often enough to give you a great view.

It is difficult sometimes to think of wild animals as shy. A Pheasant Coucal* once tried to persuade Hilary Thompson, to come into the bushes with it. When Hilary imitated its call, the bird ran up to his feet with a 'come hither' look in its eye, and then back into the bushes. When Hilary did not follow, the bird ran at him again. It persisted for about a hundred metres totally ignoring the very pregnant me and our ferocious dog. Hilary resisted the temptation to desert me for another, but only as he now says, because the coucal was 'very ugly!'. Let this attitude be a warning to those women contemplating marrying Englishmen!

Bushfires sometimes drive animals onto roads. If you see a bushfire stop and watch but do not get too near as you may panic an escaping animal which could turn and dash back into the flames.

*Michael Bollen of Wakefield Press, asks that I explain what a Pheasant Coucal is, for 'dummies like (him)'. A Pheasant Coucal is a large non-parasitic cuckoo which is not too bright as might be gathered by the above story. Incidentally the plumage of the adults is rather beautiful and I would never describe them as ugly. However the chicks are another story. They have great black gapes and white cotton-like threads sticking out of their heads.

Safety

Other humans, especially those driving motor vehicles, are the biggest health hazards to Top End travellers, more so than any other animals, except perhaps mosquitoes. Spiders, scorpions and centipedes are rarely a problem. As a precaution, keep belongings such as haversacks and shoes, off the ground or in vehicles or tents. Camp away from trees and the edges of creeks and billabongs where spiders hang out, and keep tents zipped up. If you are bushwalking without a tent then check your sleeping area for holes in the ground.

Always carry a broad bandage with you. These are handy in all sorts of emergencies including snake-bite, sprains, holding pants up if your belt breaks, and strangling babbling guides. In the case of snake, and some spider-bites, wrap the bandage firmly around the bite and up the limb (see details under snakes). Ice-packs (cold drink cans and frozen vegies will suffice) will help reduce pain and swelling in the case of bites from red-back spiders. Get to a doctor quickly. However fear can exacerbate the effect of an injury so do your utmost not to scare the wits out of your patient.

Long trousers and long-sleeved shirts will protect against cuts and scratches which may become infected. This gear, in light colours, will also make you less attractive to mosquitoes. Eating lots of Vegemite will also keep mosquitoes away, something to do with its high concentration of Vitamin B. However do not breathe over North Americans or Britishers. They hate the stuff.

Always wear a hat and closed-in footwear. A hat will protect against the sun - essential up here with our high UV ratings. (and against the blows of visitors if you eat Vegemite). Closed-in footwear will protect against most snake-bites: 97% or so occur on the lower leg. It will also protect your toes against breakage. I usually wear gumboots unless climbing rocks. They are cheap (one $30 pair lasts me four years), give knee-high protection against everything including leeches, and are great for crossing shallow creeks or muddy stretches. They also scrub up well for formal functions.

Watch out for heat exhaustion, especially during the hot, humid times of year. A slight headache or dizziness may be the first signs, or there may be no signs. Visitors from cool climates, or locals who live in air-conditioning, must be very careful. Drink plenty of water and wear loose cotton clothing. Coffee, soft drink, beer or even fruit juice are not substitutes for water and all can exacerbate heat exhaustion. If walking any distance do so slowly and stick to the shade. I've taken out Canadians in their eighties on 3-4 km walks in hot weather (In one instance they came straight from a Canadian winter). We were careful and had no problems.

Do not camp or even shelter under trees during thunderstorms. If you are standing on top of Ubirr Rock or similar places get down as quickly as possible. We have one of the highest rates of lightning strike in the world.

Early dry season bushfires are rather cool. You can even run through them if you must and suffer little more than singed eyebrows. Keep away from burnt trees, however. Most trees in woodland and open forest are hollow and can fall without warning when weakened by fire. In the late dry season fires are hotter and need to be treated with great caution. Do not camp in areas where there is long dry grass.

Safety is also mentioned at length under entries for kangaroos, snakes and crocodiles.

PARKS OF THE TOP END

Kakadu National Park, the biggest park in the Top End, covers 20 000 square kilometres of the northern region. Other major national parks are Litchfield, south of Darwin, Nitmiluk (in the southern part of the Top End), and Gregory and Keep River in the west. There are several smaller parks and reserves which are listed below by distance from major towns.

Parks and reserves in Darwin

Botanic Gardens Located along Gilruth Avenue, only a kilometre or so from the City Centre.

Casuarina Coastal Reserve Located near Darwin's northern suburbs, about 30 km from City Centre. It features monsoon and open forest, mangroves and sand dunes.

East Point Located at the end of East Point Road about 6 km from the City Centre. Habitats include monsoon vine-thicket, grasslands and rocky cliffs. Agile Wallabies and Northern Brushtail Possums are common here.

Holmes Jungle Nature Park Located about 20 km from City Centre off Vanderlin Drive on the north-east boundary of Darwin's northern suburbs. Dirt road is suitable for 2WD. It features monsoon and open forest, paperbark woodland and wetlands.

Within 70 km of Darwin

Berry Springs Nature Park Located about 55 km south-west of Darwin. Turn off Stuart Highway onto Cox Peninsula Road. It features monsoon forest and a spring-fed creek. Territory Wildlife Park is nearby with similar sorts of habitats.

Fogg Dam Reserve Located 60km from Darwin. Turn off Middle Point Road which runs off Arnhem Highway. It is sealed all the way except for the last section and the dam wall which you can drive along. Suitable for 2WD except in the late wet season. Habitats include monsoon and open forest, paperbark woodland and wetlands.

Howard Springs Nature Park Located off the Stuart Highway about 25 km south of Darwin. Habitats include monsoon and open forest and a spring-fed creek.

Manton Dam Located off the Stuart Highway about 65 km south of Darwin. There is monsoon and open forest, paperbark woodland and bamboo thickets. Some dirt road but suitable for 2WD.

Within 150 km. of Darwin

Litchfield National Park. Located about 110 km from Darwin. Dirt road for about fifty kilometres if you come in from the Cox Peninsula Road. Usually suitable for 2WD with care. The road is sealed all the way if you drive through Batchelor. Roads may be cut during wet season. Most accessible parts are crowded during mid-dry season. Habitats include monsoon forest, waterfalls, open forest and woodland and rocky outcrops and hills. Petherick's Park is in the same area and has a similar range of habitats.

Mary River Crossing Located 115 km from Darwin along the Arnhem Highway. It has rocky outcrops, bamboo thickets, monsoon forest, and a river full of large estuarine crocodiles.

Within 250 km. of Darwin

Bamboo Creek, Daly River Nature Park. Located about 250 km south-west of Darwin. Turn west off the old Stuart Highway which begins just south of Adelaide River. The Daly Raiver Road is partly sealed and usually suitable for 2WD but may be cut during wet season. A range of habitats within this area.

Butterfly Gorge, Douglas Hot Springs Located about 220 km south-west of Darwin. Turn west off the old Stuart Highway on to the Douglas-Oolloo Road, This dirt road is suitable for 2WD except possibly in the late wet season. Access to Butterfly Gorge is only during the day and only by 4WD. Areas of interest include hot springs, monsoon forest and gorges.

Kakadu National Park. There is a circular route that goes down the Stuart Highway, then the Arnhem Highway. Within the Park turn onto Kakadu Highway (partly sealed) which leads to Pine Creek. From Pine Creek the Stuart Highway runs north to Darwin. It is bout 250 km from Darwin to the town of Jabiru within the Park. Most easily accessible parts are crowded and even unpleasant during mid-dry season. There is a full range of habitats.

Further afield from Darwin.

Gurig National Park. Located about 570km. from Darwin. Turn off the Arnhem Highway near Jabiru. There is a dirt road across Arnhemland which is cut during wet season. Access is also by boat or aircraft. A permit is needed from Northern Land Council. Habitats include monsoon forest, paperbark and eucalypt woodland, mangroves and wetlands.

Pine Creek region (Pine Creek is about 210 km south of Darwin)

Umbrawarra Gorge Reserve. About 30 km from Pine Creek. Turn west off Stuart Highway. The dirt road has seventeen usually dry creek crossings. 2WD with relatively easy access except for temporary flooding in wet season. Gorge with open forest and rocky savanna woodland.

Katherine region (Katherine is about 310 km south of Darwin)

Cutta Cutta Caves Reserve. Located about 27 km south of Katherine. Turn west off Stuart Highway. Limestone caves and open woodland.

Elsey Park. 120 km south of Katherine. Turn off Stuart Highway. Dirt road is suitable for 2WD. There are limestone formations, open and monsoon forest, and the lovely Roper River.

Nitmiluk Gorge National Park. About 40 km. from Katherine. Turn east off Stuart Highway at Katherine. Walking tracks. Most accessible parts are crowded during the dry season. Features a series of large gorges with monsoon and open forest and woodland.

Further afield

Gregory National Park. 270 km. west of Katherine along the Victoria Highway. Features open woodland and rocky outcrops. This area is never crowded, in my experience, and wildlife is relatively undisturbed.

Keep River National Park. 540 km. west of Katherine on the Western Australian border. Access off Victoria Highway . Open woodland and rocky outcrops.

DESCRIPTIONS OF ANIMALS

This section begins with mammals. Reptiles, frogs and introduced animals follow on.

MAMMALS

Mammals are vertebrate animals which have fur, secrete milk and have one bone in the lower jaw. They have been in existence for about 200 million years. There are three sub-classes - Prototheria, the egg-laying monotremes; Methatheria, the marsupials, whose young are born as embryos; and Eutheria -mammals that do not fit into the other two groups.

Australia is the only continent to have all three groups extant. North and South America have both carnivorous marsupials (opossums) and eutherians. Fossils of monotremes have recently been found in South America but no living animals exist today outside of Australia and New Guinea.

THE MONOTREMES

This ancient group may have originated in the Australasian region. The discovery of an opalised jawbone of a platypus *Steropodon galmani,* 110 million years old, has given the monotremes the longest fossil record of any mammal in Australia.

There are only three living species of Monotreme, the Short-beaked Echidna, found all over Australia, Long-beaked Echidna which is now only found in New Guinea (there is fossil evidence of it in the Top End), and the Platypus, which is only found in eastern and southern Australia.

'Monotreme' means 'one opening'. This is because the intestine, bladder and the reproductive system open into a common chamber with one exit, the cloaca. Marsupials differ slightly and eutherians even more, as depicted below.

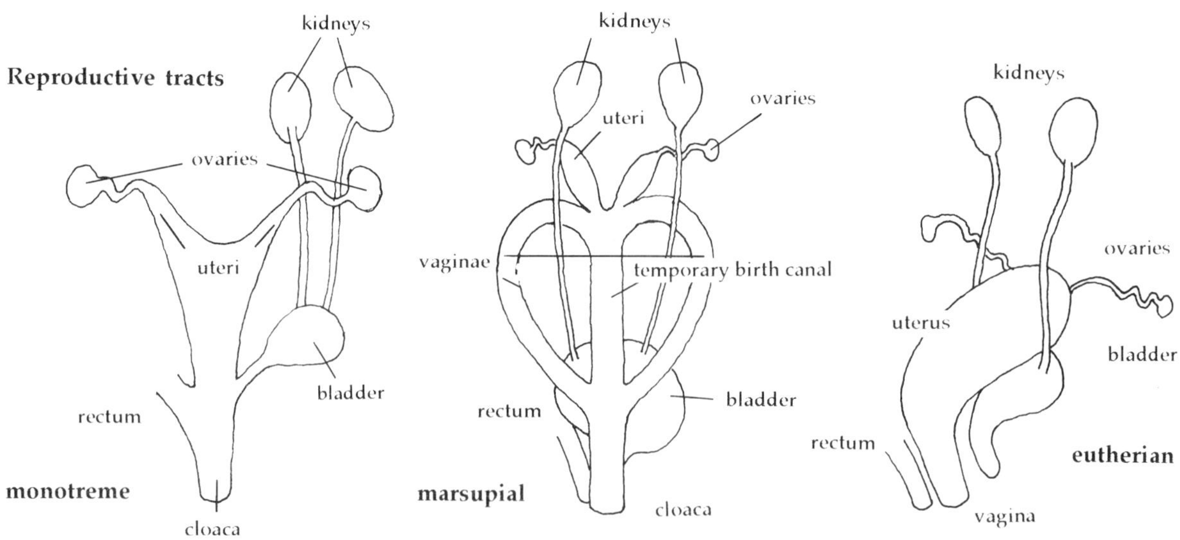

Monotremes have much in common with reptiles, and, to a lesser extent, with birds. For instance some of their skeletal parts are similar to those of lizards, and monotreme chromosomes and sperm structure are like those of birds and reptiles. Also, unlike any other mammal, monotremes lay eggs.

Yet they are definitely mammals in that they have fur, feed their young on milk and have only one bone in the lower jaw. Also, unlike reptiles the monotreme brain is relatively large with a well-developed neocortex. This suggests that like primates, bats and whales, monotremes may have a large capacity for problem-solving.

Other interesting anatomical information
Unlike 'placental' mammals the female echidna has two uteri and vaginae (see picture, page 16), a feature she shares with marsupials. The male echidna's penis has a rather distinctive appearance resembling a miniature cauliflower on a stalk more than anything else (see picture below). As with marsupials the penis is held internally.

The testes are within the abdomen. In most mammals the testes descend at maturity because higher body temperatures lower the viability of the sperm (men who wear tight jeans undo this master plan by raising their scrotal temperature, often rendering themselves sterile). However monotremes have a low body temperature which does not affect their sperm. Elephants, dugongs and whales also retain the testes within the abdominal cavity.

Male monotremes have a spur on the hind ankle. In the platypus, but not the echidna, this is attached to a functional venom gland. According to Dr R. Martin and G. de Plater of the Australian National University the platypus sting can causes excruciating pain, swelling and paralysis which can last for years. However it does not seem to have killed anybody.

ECHIDNA *(Tachyglossus aculeatus)*

Description: Spiny long-nosed animal with long thick claws on its hindfeet. 35 cm. Cat-size. Common but relatively hard to see in the Top End.

Habits: Mainly nocturnal in the Top End, unlike down south where it is diurnal. It feeds on termites and ants including meat ants, *Iridomyrmex* species, which extracts from nests with its tongue which is covered with sticky saliva. It is usually a solitary animal except in the breeding season. However they can form colonies in suitable habitat.

Echidna

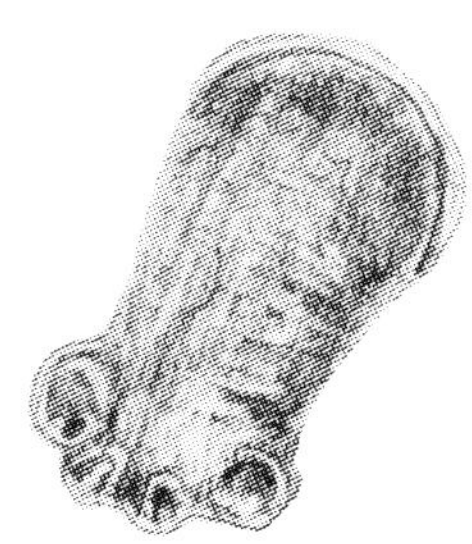

Echidna penis

Mating behaviour
Most people are curious about how such a prickly animal mates. Read on.

Male echidnas form 'trains' of two to twelve individuals prior to mating. Dr. Peggy Rismuller, who researches echidnas at Pelican Point on Kangaroo Island, reports repeated sightings of males digging a trench around a female, who lies on her stomach with front feet anchored, eg hanging onto the base of a tree. The males shove each other, head on head, out of the 'mating rut' until only one remains. He continues to dig along one side of the female, stroking her spines with his forefoot, to which she responds by 'undulating' her quills. He attempts to lift her posterior and, when he has succeeded, lies on his side in the trench with his tail under hers. Mating takes place with cloacas touching. It can also take place with the echidnas facing in opposite directions. Dr. Rismuller notes that the echidnas remained together like this for between one and three hours.

Incubation and raising of young
Females lay a single egg which is deposited in the pouch to incubate. According to Dr. Peggy Rismuller no one has seen how the eggs get there *(Scientific American Feb. 1991)*. Echidna researchers suggest that the echidna's extruded cloaca could reach into the pouch.

Dr. Mervyn Griffiths determined that an echidna egg hatches after about ten days of incubation. The young, or 'puggle', has an egg tooth which enables it to break through the eggshell when hatching (birds and reptiles have an egg tooth as well). The puggle also has well-developed fore-limbs which are used to hold onto hairs in the mother's pouch, as the female's milk ducts do not form nipples that it can grip. To feed, the young simply sucks milk from specialised pores at a spot called the 'milk patch'. It lives for about sixty days in a temporary pouch, which is only well-formed while carrying the baby. When the puggle grows too big for the pouch it is deposited in a nursery burrow.

Where found: In rocky outcrops, woodland and open forest all over Australia and in New Guinea. There is a record of one being run over on Dick Ward Drive, Darwin *(Northern Territory News*, August, 1980) and others from Palmerston (no date). Records from most Top End national parks and Manton Dam and Umbrawarra Gorge Reserves. Also Batchelor, Douglas, Daly and Mary Rivers, Noonamah, Oenpelli and Pine Creek.

Notes: 'Echidna' comes from the Greek word 'echinos' meaning spears. *'Tachyglossus'* is Greek for 'swift tongue'. Gunwinggu people from Western Arnhemland call it Ngarrobek and regard it as 'good tucker' or even as medicine. They hunt Ngarrobek at night. It is gutted and filled with hot stones and bush herbs, namely the leaves of Manduk *(Persoonia falcata)*.

YOU CAN HELP! Dr. Rismiller's long-term studies on Kangaroo Island are following individual echidnas from egg to adult. Living amongst a wild study population makes it possible to observe and thoroughly document behaviour and population dynamics as never previously done. As one of the oldest surviving mammals in the world, the echidna has many unusual and successful traits. By discovering how and why echidnas have been so successful we are gaining new insights into the biology of life. As global options become fewer, many species will benefit from the strategies of the survivors.

Echidnas are generally solitary, secretive and not easily counted. ECHIDNA WATCH is the first attempt at a nation-wide Echidna survey. Data will be collected for many years. Have you seen an echidna? Both current and historical sighting are important to help piece together The Echidna Story. Write down your Echidna encounters or copy the survey form in the back of the book. Send to ECHIDNA CARE INC, 8 Edward Avenue, Hawtorn SA 5062. Every observation helps!

MARSUPIALS

Marsupials are another ancient family of mammals. North American fossils, about 100 million years old have been discovered. They are also known from Europe, South America and Antarctica.

The earliest Australian marsupial fossils are only 55 million years old. However marsupials diversified more here, because of the absence of competition from 'placental' mammals. These, apart from bats, probably did not enter Australia until 15 million years ago.

Marsupials differ from 'placental' mammals in a number of ways, the most obvious being that marsupials have a pouch (the term 'marsupial' comes from the Greek *marsipos,* meaning bag or pouch). Also marsupial young are born as embryos (see below), while those of 'placental' mammals are born fully-formed.

Marsupial females are equipped with two vaginae. Sperm ascend either of these. However the vaginae are not involved in birth which occurs through a third structure, the median vagina, or temporary birth canal (see picture, page 16). This latter structure is open only temporarily in many marsupials and closes after birth.

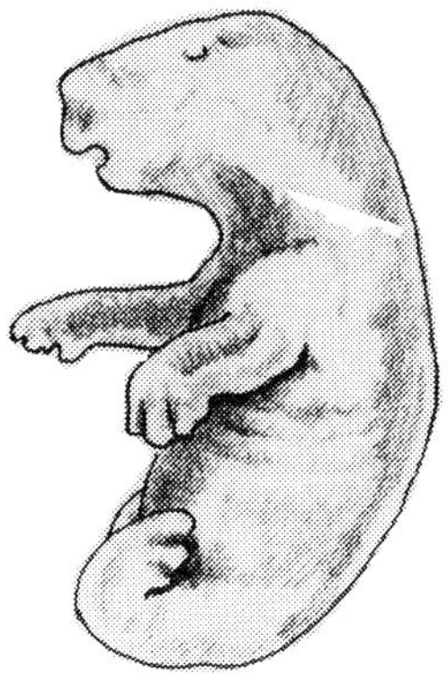

Marsupial young at birth

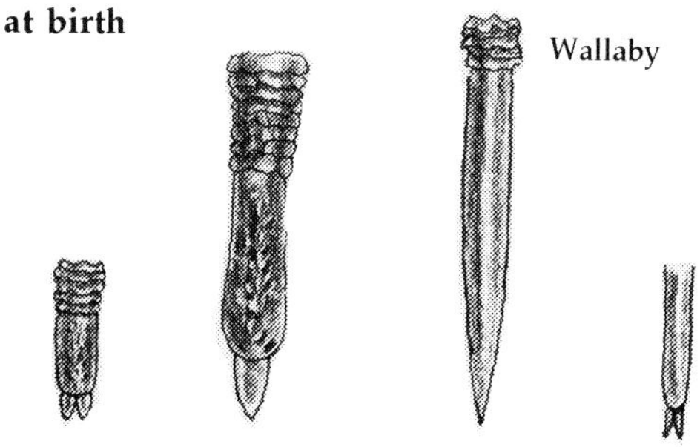

Penes

Dunnart

Brushtail Possum

Wallaby

Bandicoot

The penis varies between different groups of marsupials (see above). Many marsupials have a split or forked glans penis often equipped with backward-pointing spines. Kangaroos have a long whip-like affair. The marsupial male's testicles are external and, unlike most other mammals, except some rodents and rabbits, are above the cloaca. When the penis is not in use, it is folded into an 'S' shape within the cloacal opening.

Some marsupial prostate glands are large during the breeding season, when they become the second largest organ in the body after the liver.

Many carnivorous marsupials do not have a pouch, but drag their young behind them until they are old enough to be left on their own. Other marsupials have temporary pouches. Some, like bandicoots, wombats and koalas, have backward-opening pouches. Others such as kangaroos and possums have forward-opening pouches. Both male and female marsupials and monotremes have epipubic bones attached to the pelvis. The role of these bones is not clear but they possibly support the pouch muscles. Many marsupials do not have a pouch but they all have epipubic bones.

Marsupials are divided into two orders, Polyprotodonta - the carnivorous marsupials, and Diprotodonta - the herbivorous marsupials. Polyprotodonta includes bandicoots, so-called native cats or quolls, and marsupial-mice. Diprotodonta includes the possums, kangaroos, koalas and wombats -the latter two groups are not found in the Northern Territory.

THE CARNIVOROUS MARSUPIALS
Order: Polyprotodonta

This order is represented in the Top End by the family Dasyuridae. 'Dasyurus' means 'hairy-tail'. Another group, represented by the Tasmanian Tiger, or Thylacine, is apparently extinct. (See Thylacine, page 42).

Many species of small carnivorous marsupial look mouse- or rat-like. However their teeth differ greatly from those of rodents and the herbivorous marsupials, in that they have four or five pairs of pointy upper incisors and three pairs of lower incisors. Their fore-feet look much the same as their hindfeet (unlike kangaroos and bandicoots) and they all have five digits on their forefeet, unlike rodents which have four digits, the thumb being reduced. Many do not have a permanent pouch. Instead flaps of protective skin develop in the pouch area.

Mating behaviour among many dasyurids is not subtle, the males being quite savage and sometimes causing serious injury to females. In some species the males do not survive to mate a second time which, as far as I'm concerned, serves them right. They seem to expend so much energy that they become vulnerable to disease and predators.

Copulation may last for several hours, penetration usually taking place from the rear with the female's tail moved to one side. One South American dasyurid cannot achieve erection unless it has its tail wound around a branch or similar object. I know of no Australian dasyurid, or any other male for that matter, that has a similar requirement.

NORTHERN QUOLL ***(Dasyurus hallucatus)***

Description: Solitary brown spotted nocturnal animal with long unspotted tail. Large rat-size. 30 cm (excluding tail). There is no true pouch, just a temporary arrangement consisting of two longitudinal flaps of skin surrounded by long hairs. Unlike most dasyurids it has five toes.

Habits: An aggressive animal that hisses loudly if cornered. It feeds on frogs, worms, other mammals and fruits. Males, which are twice as big as females, will defend a territory. Young are born in June-July and there are often many more young than nipples (nipple number varies from six to eight). Females do not breed until nearly a year old.

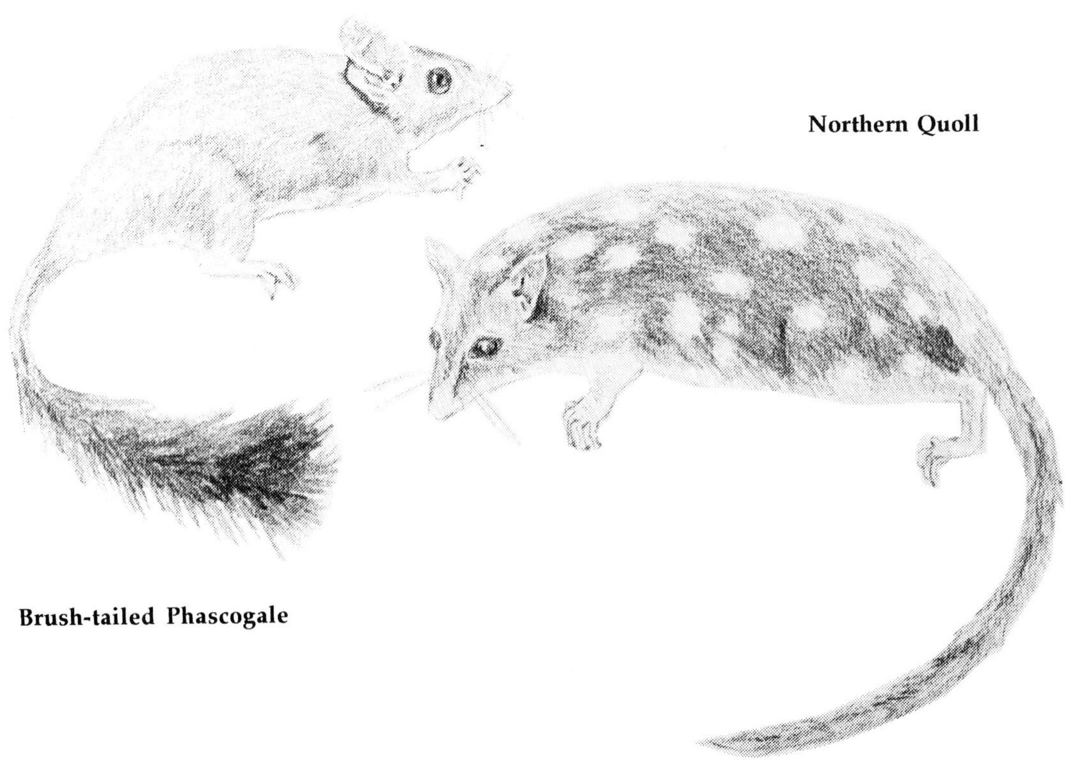

Northern Quoll

Brush-tailed Phascogale

Where found: Found right across the coast of northern Australia in rocky outcrops and open forest. In surveys carried out in the Pine Creek area I found it to be the most commonly trapped animal. Top End records from Darwin (Holmes Jungle) and Fogg Dam, Manton Dam, Howard Springs and Umbrawarra Gorge Reserves and Gurig, Kakadu, Litchfield and Nitmiluk National Parks. Often seen on roads at night eg along the Stuart and Arnhem Highways.

Notes: This animal is also called, quite wrongly 'Native Cat' (true cats are eutherians).

Gunwinggu people call it Djabo and say it is 'good tucker.' The hair is singed off, the guts are removed (except for the heart and liver) and the cavity packed with bush herbs. It is roasted on coals in a hole.

My friend Jan came across a Northern Quoll sitting on the floor of a washroom at the Frontier Lodge in Jabiru. Judging by Jan's excitement at seeing this animal at such close quarters, it was the highlight of her trip to Kakadu.

BRUSH-TAILED PHASCOGALE *(Phascogale tapoatafa)*

Description: Solitary brown/grey animal with a long black , bushy tail like a bottlebrush, big ears and a rat-like face. Rat-size. 20 cm (excluding tail). (See picture previous page).

Habits: This arboreal animal lives in tree hollows in tall, open forest. and feeds on spiders, centipedes and other invertebrates, and small vertebrates such as birds and lizards. When alarmed it taps on trees, probably to alert other animals, and erects the hairs on its tail.

The hindfoot can be rotated allowing the Phascogale to climb up or down. It also has an opposing first toe, like the possums. The female, like the Northern Quoll, lacks a permanent pouch. She produces up to eight young which attach to the nipples and are dragged behind the mother until old enough to be left in a nest.*

Where found: Across northern Australia and down the east coast, with a small distribution in southern Western Australia. They are found throughout the northern Top End with records from Gurig and Kakadu National Parks and from Borroloola. Phascogales can be seen at night scampering across roads or as roadkills. These fatalities are mainly males. I often see them along the Stuart , Arnhem Highway and Kakadu Highways.

Notes: *'Phascogale'* means 'pouched weasel' in Greek.

Similar species: See **Brush-tailed Rabbit-rat** *(Conilurus penicillatus)*

*Many carnivorous marsupials, and rodents, drag their young behind in this way. The young of some rodents have specialised, incurved incisors that allow them to grip the teats (women such as myself who are familiar with the sore nipples that often come with breastfeeding shake their heads in amazement at this feat). Older young run behind the mother while still holding onto the teats.

FAWN ANTECHINUS *(Antechinus bellus)*

Description: Pale grey, rat-like nocturnal animal with a pointed face and large ears which inhabits the understorey of open forest, often near ceeks. Underparts and feet are white. Small rat-size. 12 cm. (excluding tail). Relatively common but not a highly visible animal. (Picture next page).

Habits: Fawn Antechinus eats insects and spiders. It breeds during August-September after which the males die. The females have a variable number of nipples (6-10) like other Antechinuses. They bear big litters (up to ten) and suckle for up to four months. Females of this genus lack a true pouch.

Where found: In gallery and eucalypt forest near creeks across the Top End. Records from Gurig, Kakadu and Litchfield National Parks and the Mary River area.

Notes: Antechinuses are also known as broad-footed marsupial-mice. ('marsupial-mouse' is a common name for small carnivorous marsupials). This animal and Sandstone Antechinus (below) are called Mulbu by Gunwinngu people, a name also given to many rats.

Similar species: There are some rodents of similar colouration and size that inhabit similar areas eg Grassland Melomys (page 38) and Pale Field-rat (page 40). They are very difficult to tell apart unless in the hand. However Fawn Antechinus has a much more extensive earflap than the others, it is greyer and its tail is furred. *Melomys burtoni* and *R. tunneyi* are usually browner and have a bare tail (see under these names in section on rodents).

SANDSTONE ANTECHINUS *(Antechinus bilarni)*

Description: A small grizzled brown mouse-like animal with a pointed snout, found in sandstone. Mouse-size 8-10 cm. Relatively common in a limited area.

Habits: Sandstone Antechinus eats mainly insects. It has up to six young at a time, usually around August to September, but rarely all survive. Unlike many dasyurids, male Sandstone Antechinus may survive after mating, sometimes into a third year.

Where found: Records from sandstone escarpment within Kakadu, Litchfield and Nitmiluk National Parks and the Borroloola region.

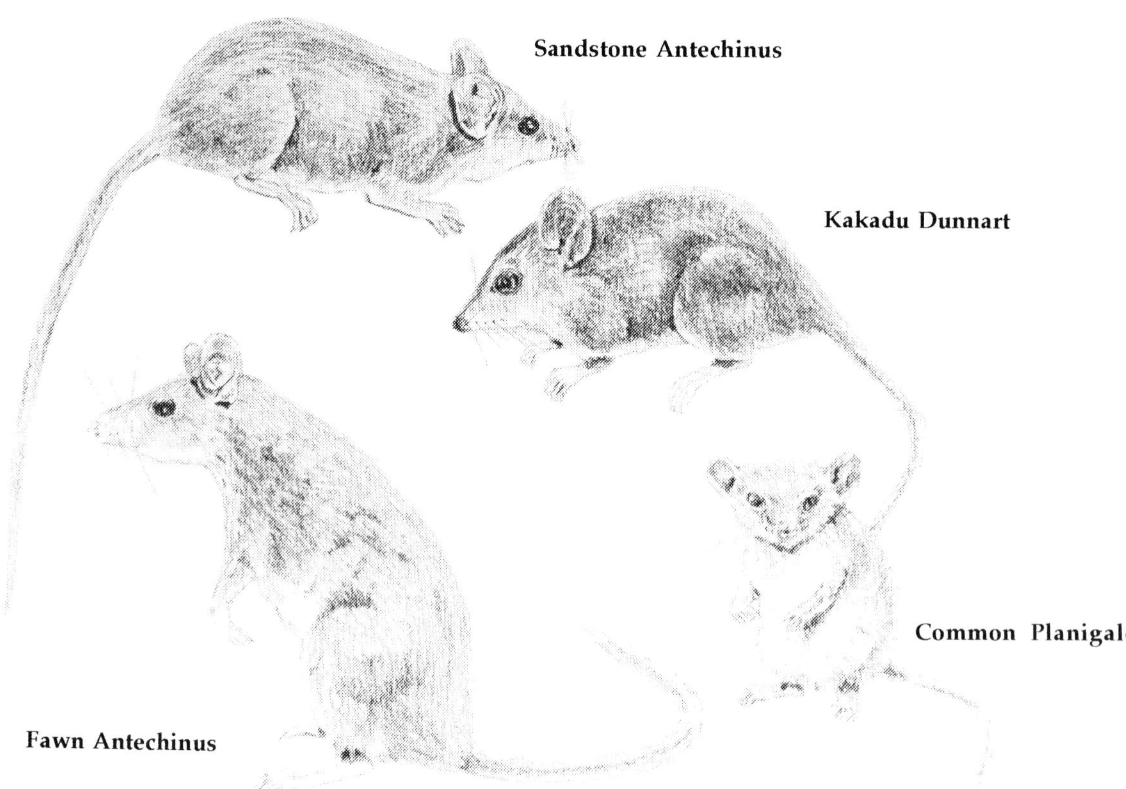

KAKADU DUNNART *(Sminthopsis sp).*

Description: Solitary. brown-grey, mouse-like animal with a stripe down its face. Mouse-size. 9-10 cm (excluding tail). (See picture previous page).

Habits: Dunnarts eat invertebrates and small vertebrates. Females have a well-developed pouch and may have two litters of up to eight young in the one season. Older young are left in a nest. Female dunnarts attack and even kill males when they have pouch young.

Where found: Kakadu Dunnart inhabits savanna woodland on stony hills and has been recorded from southeastern Kakadu National Park and the Pine Creek-Katherine region. Dr. Richard Noske reports unidentified dunnarts similar to this species trapped near Batchelor.

Notes: Martin Schultz and I caught two Kakadu Dunnart in pitfall traps while working at the Mt Todd mine site in late 1989. The remains of a third were found at a Ghost Bat roost in an old adit in the same area. These, and further records, refute the claim made in the Kakadu Stage 3 wildlife survey report (CSIRO, 1989) that they are found 'almost exclusively within Stage III and the Conservation Zone (this is now part of Kakadu Stage III).

The soles of the feet of different species of dunnart differ. This is an important aid to identification for scientists. Dunnarts are also called narrow-footed marsupial-mice .

RED-CHEEKED DUNNART *(Sminthopsis virginiae)*

Description: A grizzled grey-brown mouse-like animal with rust-red fur on its cheeks. Common in limited areas.

Habits: See Kakadu Dunnart above.

Where found: Red-cheeked Dunnarts inhabit savanna woodland and grasslands that may be seasonally inundated across northwest Australia, eastern Queensland, and in New Guinea. Records from southeastern Kakadu National Park and Mt Todd.

COMMON PLANIGALE *(Planigale maculata)*

Description: A tiny grey/brown animal with a very broad, flat, triangular head and a thin tail which is shorter than the head and body. Smaller than a mouse. 7-8 cm (excluding tail). Common but not very visible. (See picture previous page).

Habits: A ferocious hunter, this animal tackles invertebrates almost its own size. Males also fight savagely among themselves. The female has up to twelve young which she carries in a well-developed rear-opening pouch. Older young are left in a nest made of bark and grass. Planigales often shelter in the deep cracks of the floodplains during the dry season. When the rains begin the clay swells, closing the cracks and forcing these animals into the open where they are picked off by owls and other predators.

Where found: In a range of habitats, including eucalypt woodland, in the northern Top End and down the east coast of Australia. Records from Holmes Jungle, Fogg Dam Reserve and every major national park. Also Mary River, Borroloola and Pine Creek regions.

Notes: Planigales are among the smallest mammals in the world. *'Planigale'* means 'flat weasel' in Greek, a very appropriate name. This animal's head is so flat that the first time I caught one I thought it had met with a nasty accident that had crushed its skull!

Similar species: Long-tailed Planigale, *P. ingrami*, is found in semi-arid regions aross most of northern Australia. It has been recorded from Gregory , Kakadu and Keep River National Parks and the Borroloola region. Smaller even (6 cm excluding tail) than Common Planigale, with a longer tail. It is the second smallest mammal in the world being just beaten by the African Shrew.

THE BANDICOOTS
Suborder: Peramelomorphia

Bandicoots have hindlegs similar to those found in kangaroos. The foot is elongated, the first toe is reduced, the second and third toes are syndactylous (joined together)* and the fourth toe is large with a big claw. Yet bandicoots have a dasyurid-like dentition, with many incisors. Like both koalas and wombats bandicoot pouches have a rear opening.

The related bilbies were once found over much of Australia but are now confined to central Australia. Two species of bandicoot are recorded for the Top End.

*Syndactylous toes are used for grooming.

NORTHERN BROWN BANDICOOT *(Isoodon macrourus)*

Description: A solitary, dark brown, bulky animal with a pointed nose and a thin, short tail. Cat-sized. 40 cm (excluding tail). Males are larger than females. Common.

Habits: This nocturnal animal is omnivorous, eating fruit and invertebrates. After years of having campsites raided I conclude that they prefer a diet similar to that of the teenagers I know. Bandicoots especially like that abominable spread, peanut butter and jelly.

The day is spent in a rough nest of grass or leaf litter. They are solitary animals, often aggressive to others. However they can become half-tame. One came into Hilary Thompson's camp to drink water from a pan at his feet.

Males have a scent gland behind the ear which they use to mark territories. Females have several litters of up to six young a year.

Where found: This species is found in open and gallery forest across most of northern Australia, down the east coast, and in New Guinea, . It is found throughout the Top End with occasional sightings in Darwin, along Dick Ward Drive, at Casuarina Coastal Reserve and Holmes Jungle. Other records from Fogg Dam, Manton Dam, and Mary River Crossing Reserves, all the major Top End national parks, and the Pine Creek and Borroloola regions.

Related species: The endangered Golden Bandicoot, *Isoodon aurutus*, is known from Kakadu National Park and northwestern Australia. This animal has a coarse coat of golden-brown and is found in a range of habitats.

Notes: 'Isoodon' means 'equal-toothed' in Greek. The word 'bandicoot' is actually the name of a large rat, *Nesokia* species, that is found in India. Bandicoot is called Yok by Gunwinngu people (big bandicoot is Ngunma Yok), and is an important food. It is stuffed with hot coals and bush herbs, covered with paperbark and cooked in a ground oven.

Northern Brown Bandicoot, and another species, *Perameles nasuta,* has the shortest gestation period of any mammal - 12.5 days. To give birth female bandicoots lie on their sides with a leg raised. The young are well-equipped for the epic journey to the pouch, with clawed forefeet (the claws are shed after birth). Bandicoots, unlike other marsupials, have a true placenta.

Northern Brown Bandicoot

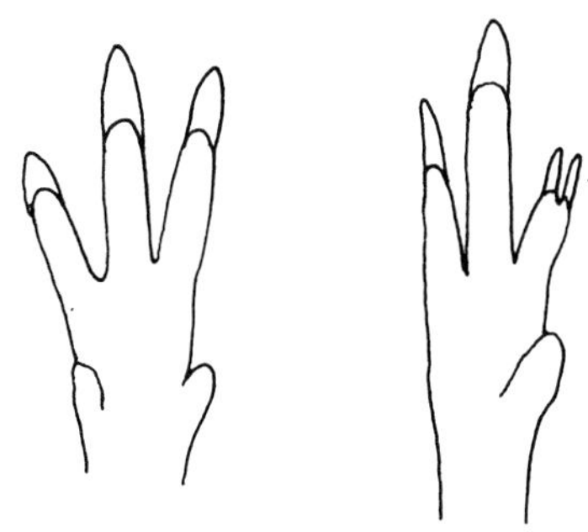

THE HERBIVOROUS MARSUPIALS
Order: Diprotodonta

This order contains koalas and wombats (which are not found in the Top End), the possums, kangaroos, and extinct groups such as Diprotodons. One of these *D. optatum,* was bigger than a cow. This huge creature only became extinct about 20 000 years ago, and would have co-existed with early Aboriginal people.

Diprotodonts are basically distinguished from polyprotodonts by their teeth (diprotodonts have one pair of lower incisors) and their largely herbivorous diet.

THE POSSUMS
Super-family Phalangeroidea

These animals have opposed first and second digits on their forefeet and an opposed hallux (big toe) on the hindfeet. These adaptations help them grip. The second and third toes are syndactylous, an arrangment also found in bandicoots and kangaroos. Females have a forward-opening pouch. There are two species of possum and one glider in the Top End. All are nocturnal.

ROCK RINGTAIL POSSUM *(Pseudocheirus dahli)*

Description: A bulky grey or reddish, rock-dwelling possum with thick fur and a dark stripe down its back. Tip of its short tail is almost naked. Cat-size. 35-38 cm (excluding tail). It may be solitary or in small groups. Common in limited areas. (Picture next page).

Habits: This animal is restricted to rocky habitats although it is occasionally found in nearby trees. It feeds on fruits eg. *Owenia vernicosa* and *Ficus platypoda.* Like other ringtail possums both the first and second digits are opposed on the forefoot. It usually has one young at a time, usually in the dry season. I have noted well-grown young in late June.

Where found: in sandstone habitat in the Top End and northern Western Australia. Recorded from

Kakadu, Keep River, Litchfield and Nitmiluk National Parks, and Manton Dam and Umbrawarra Gorge Reserves. Also Douglas River, Pine Creek and Borroloola region. There are six members of this genus in Australia and eight in New Guinea.

Notes: Low-powered torches held to one side do not seem to frighten these animals if carefully used. **DO NOT** shine spotlights in its eyes or in the eyes of any other animal for that matter. Careless spotlighting can cause these possums to desert an area.

Possums have pink eye-shine like many other vertebrates, including fish, crocodiles and snakes. This is caused by a layer behind the retina called the tapetum lucidum which reflects light. Gunwinggu people hunted this animal which they call Ngingma, by placing sugar bag (bush honey) on rocks. The animal was attracted to the bait and then speared.

SUGAR GLIDER *(Petaurus breviceps)*

Description: Pretty tree-dwelling animal with velvety grey fur and a dark dorsal stripe extending from its nose down its back. There is a membrane between fore and hind feet. Rat-size. 20 cm (excluding tail). Only the first digit of both fore and hind feet is opposed. Common but hard to see.

Habits: A nocturnal possum that spends the day curled up in a hole in the trunk with several of its brethren. The group comprises females and their young and a dominant male. Two young are born at a time and several litters may be produced each year. It feeds on *Terminalia grandiflora* and paperbark blossoms, on sap and invertebrates, gliding from tree to tree. They are not terribly wary - I have seen them at night feeding within two metres of my chair. Its ear-splitting scream can frighten the wits out of unwary campers.

Where found: In eucalypt forest and woodland, although more often heard than seen. Found right across Northern Australia and in a broad band down the east coast to Victoria. Top End records from the major national parks and also Umbrawarra Gorge Reserve. Also from Douglas and Mary Rivers, and the Pine Creek and Borroloola regions.

Notes: called Lumbalk by Gunwinggu people, who used to eat them.

NORTHERN BRUSHTAIL POSSUM *(Trichosaurus arnhemensis)*

Description: Grey/brown sometimes reddish, tree-dwelling animal with a long sparsely furred tail. Cat-size. 40 cm (excluding tail). Solitary and common. (Picture previous page).

Habits: This animal feeds on blossom and fruit, including the poisonous tree, Cooktown Ironwood, *Erythrophloeum chlorostachys.* This tree is well known for poisoning livestock but the Northern Brushtail seems to come to no harm. Although normally a solitary animal, immatures appear sociable. A delightful, young female shared our bathroom cabinet for a while. They usually have only one young.

Where found: Found across northwestern Australia. It is resident in suburban gardens in Darwin, at the Yacht Club, the Transit Centre in Mitchell Street, Darwin, at East Point and the Botanic Gardens. Recorded from Kakadu National Park and the Mary River wetlands, Bamboo Creek, Daly River, and the Borroloola region. They seem to be scarce in some areas, and, ccording to locals this is because of past hunting by Aboriginal people. Gunwinggu people calledthis possum Djabow and say it 'tastes good'. It is cooked in a hole like bandicoot.

KANGAROOS
Family: Macropodidae

'Macropod' means 'big foot' in Greek. Macropodids have syndactylous toes, like bandicoots and possums. A wallaby is a small kangaroo and a wallaroo, a rock-dwelling kangaroo.

Several million years ago Australia was covered with rainforest. It gradually became drier about 10-15 million years ago, and open forest and savanna woodland began to dominate the landscape. Kangaroos were probably arboreal, rainforest animals, originally. With the spread of grasslands, grass-eating and browsing kangaroos like many of those found today evolved.

The ranges of many species of kangaroo have changed both before and since European settlement. Before, such changes were possibly caused by increasing aridity and predation by Aboriginal people. Since European settlement causes have mainly been loss of habitat and predation by introduced animals. Some species such as the elegant Toolache and the Crescent Nailtail Wallaby, have probably become extinct. Others, such as the Mala of Central Australia, are under threat. However other species, eg Red and Grey Kangaroos have increased their range and numbers through factors such as farming in marginal areas which supplies them with food and water. These changes have had less effect on wildlife in the Top End.

The female kangaroo can ovulate only a few days after birth - it is not suppressed by the fact that she is feeding young. The fertilized egg will cease development at the blastocyst stage, until young are almost ready to leave the pouch. This phenomenon is known as embyronic diapause, possibly a very ancient feature of mammalian reproduction. The pouch, unlike that found in most polyprotodonts, is permanent.

Kangaroos either lean back on their tails or against a tree while giving birth. They have four teats which can supply different formulae to young of different ages. Some eutherians can also do this, eg giraffe, elephant, rabbit.

NORTHERN NAILTAIL WALLABY *(Onychogalea unguifera)*

Description: A solitary, sandy-grey kangaroo with big ears, a dark stripe down its back, pale shoulder and flank stripes, and a horny nail on the tip of its tail, the latter half of which is black. This animal appears rather Roman-nosed with a square muzzle similar to that of a male Antilopine Wallaroo. Medium-dog size. 0.75 m. (excluding tail). Relatively common in some areas.

Habits: Feeds on grass, herbs and fruit. They are sometimes known as 'organ-grinders' because they move their front feet in a circular motion as they run.

Where found: sparsely distributed across northern Australia and the southern Top End in floodplains, savanna woodland and rocky outcrops. Records from Kakadu and Keep River National Parks. Hilary Thompson and I have seen roadkills near Beatrice Hill, the Carpentaria Highway, Mary River and East Baines River. Hilary Thompson also has a record from the Edith Falls Road in Nitmiluk National Park. Also known from Daly River, Fountainhead and the Borroloola region.

Notes: There were three species of Nailtail Wallaby in Australia. One species, Crescent Nailtail is extinct. The range of another, the Bridled, has declined since white settlement. However Northern Nailtail seems relatively secure at the moment. People at Mataranka rescued an orphaned Nailtail joey which Hilary Thompson and I saw. It was not gregarious like other joeys they had kept, eg Antilopine Wallaroo and Agile Wallaby.

Similar species: Northern Nailtail is the same size as a female Agile Wallaby. Antilopine Wallaroo and Common Wallaroo are much larger. However all three lack the distinctive dorsal stripe.

COMMON WALLAROO/EURO *(Macropus robustus)*

Description: A solitary grey/brown, heavily built, shaggy kangaroo, paler underneath. Dog-size. 0.6 -1.2 m. (excluding tail), males being much bigger than females. Feet and tip of tail are dark. It hops in an upright position with forepaws close to the body. Relatively common in some areas.

Habit: A solitary kangaroo which can manage well without permanent water and on poor grasses as long as it has shelter. It becomes sexually mature at 18-24 months.

Where found: Throughout much of Australia in rocky outcrops, hills and plains. More common in the southern Top End than the north. Records from Umbrawarra Gorge Reserve, and all the major Top End national parks. Also in the Borroloola and Pine Creek regions.

Notes: Gunwinngu people call this species Ngabudj ('j' is lazy and is pronounced as 'je'). They also have separate names for male - Gaklbard, and female - Wallaar.

Similar species: Antilopine Wallaroo ***(Macropus antilopinus)*** Both sexes are reddish or sandy, and white underneath. Female has a grey head and neck, or may blue-grey all over. The coat is not shaggy. Antilopines are more slender than Common Wallaroo but can grow slightly larger. They are often seen in groups.

Antilopine Wallaroos inhabit eucalypt forest and hilly areas, and are common throughout much of the coastal and sub-coastal areas of northern Australia. Records from Darwin, Fogg Dam, Manton Dam and Umbrawarra Gorge Reserves and all the major national parks. Also records from Borroloola, Fog Bay, Mary River, and the Douglas River/Pine Creek area. Gunwinggu people call the male Golobora and the female Gandi.

WARNING: Large kangaroos can be dangerous. They have been known to disembowel dogs and male rivals, using the long-clawed hindlegs in a punching action. My small daughter and I were once attacked at an animal park by a large kangaroo which grabbed first her , and then me in its its forepaws. I punched it hard in the nose before it was able to bring its hind legs up. Then we fled. Stay away from large kangaroos in zoos, wildlife and caravan parks and anywhere else where they roam free.

BLACK WALLAROO *(Macropus bernardus)*

Description: A dark grey/black animal occasionally glimpsed racing up the side of the Arnhemland escarpment in the early morning. Dog size. 0.75 m. (picture previous page).

Habits: Relatively little is known about this species. They are very wary and solitary animals. Aboriginal people who have kept them as pets say they are 'quiet'.

Where found: Restricted to the Arnhemland escarpment. Have records from Nourlangie and Ubirr Rocks, from Gunlom (Waterfall) Creek in Kakadu National Park and from the Oenpelli area but are probably found all through the area. I surprised one which sprang up almost from beneath my feet at the top of Gunlom Falls. It is called Nadjanam by Gunwinggu people.

SHORT-EARED ROCK-WALLABY *(Petrogale brachyotis)*

Description: A small grey/brown animal that lives in rocky areas. It has pale underparts, a pale cheek stripe, a pale and a dark shoulder stripe, and a partial dorsal stripe. The tail is carried in a raised position unlike that of other kangaroos. Small dog size (0.5 m). Common in suitable habitat. (See illustration page 28).

Habits: This little animal hides in caves and on rock ledges throughout the day, emerging to feed before dusk and returning to shelter after sunrise. Relatively common but not easily seen. Early morning visitors at places like Robin Falls may see them disappearing up the hillsides. I have sometimes noticed them peering out of caves at visitors.

Where found: Found in suitable habitat across northwestern Australia. Top End records from Gregory, Kakadu, Keep River and Litchfield National Parks and Butterfly Gorge and Umbrawarra Gorge Reserves. Records also from Mary River and Robin Falls, Borroloola region and throughout Arnhemland.

Notes: Called Botbong by Gunwinngu people who used to trap it in the rocks by setting the spinifex on fire (actually spinifex is an inaccurate name. The only true spinifex in the Top End is *Spinifex longifolius*, a grass which grows on coastal dunes. The grasses on the escarpment are actually *Plectrachne* or *Triodia* species. They burn well when set alight because they have a high resin content).

Similar species: Nabarlek ***(Peradorcas concinna)*** also lives in rocky country. It is much smaller (cat-size) and is grey/brown with rufus highlights on the rump, a dark shoulder-stripe and a dark tip to its tail. 'Nabarlek' is a Gunwinggu name. Uncommon. (See illustration, page 28)

Unlike other macropodids it produces molars continually, probably because its diet includes ferns and grasses with a high silica content which wear its back teeth down. Found across the Top End and in northern Western Australia. Top End records from Kakadu, Gregory and Litchfield National Parks and Umbrawarra Gorge Reserve. Also Robin Falls, Mary and Victoria Rivers and Arnhemland.

AGILE WALLABY *(Macropus agilis)*

Description: A sandy wallaby with white underparts and a pale cheek and flank stripe. It has an upright stance. A sociable animal, it is often found in large groups in open and monsoon forest. Medium-dog-size (0.5-0.7 m). Common. (See picture page 28).

Habit: An animal of catholic tastes, it both grazes on grasses, browses on leaves and native fruits and will dig for roots. As with many kangaroos, males are larger than females. Females reach sexual maturity at about 12 months, males at about 14 months.

Where found: Found in riverine habitat, open forest, paperbark woodland and monsoon forest with an open understorey right across Northern Australia, and down the east coast of Queensland. This species is common around Darwin, reaching plague proportions at East Point where young have been reported as dying of starvation (personal comment made to me by a veterinary surgeon). They are also common at Holmes Jungle. Records from every major Top End national park, Fogg Dam, Howard Dam, Manton Dam and Umbrawarra Gorge Reserves, and Daly and Mary Rivers, and Pine Creek and Borroloola regions.

Notes: We have reared orphans, finding the young affectionate and easy to handle although adults, especially males, may be aggressive and should not be kept. Our two year old son used to sit feeding himself and a joey from bottle held in each hand. The young make a raspy cough when nervous.

Young kangaroos, like the young of many other animals have an intolerance for the lactose found in cows' milk. Apparently it forms deposits behind the eyes blinding them. It also affects their digestive system and they can die. An artificial formula made of two scoops of milk powder (low lactose), 100 ml of water, one egg yolk (rinse under tap to remove white), two tablespoons of cream, and a scoop of high-protein cereal, is suitable for furred joeys. For advice on younger ones contact Berry Springs Wildlife Park.

I have eaten this animal with Aboriginal people and it is very good. Gunwinggu people call Agile Wallaby, Gongabolor, the male Warradjukal and the female Malbar.

EUTHERIANS/'PLACENTAL' MAMMALS

All the following mammals are eutherians or 'placental' mammals.

'Eu' is a prefix of Greek origin meaning 'good'. Traditionally eutherians were considered to be the most highly evolved animals in the same way that Caucasians were considered to be more advanced than other races, and men were thought to be superior to women. Today many of us, though not all, know better.

The term "'placental' mammal" is definitely inaccurate because it suggests that only these animals have placentas. However, some marsupials (bandicoots) also have true placentas.

BATS
Order Chiroptera

Bats are the only mammals that can truly fly, although some can glide. There are 950 species of bat in the world and about sixty in Australia. They are divided into two suborders. The Megachiroptera ('large hand-winged' in Greek) are large, with claws on the first and second digits of the wings. They have a short digestive tract, feeding mainly on soft fruit and nectar, although some also take insects. Some Megachiroptera can echo locate* in a rudimentary fashion. but most navigate by sight alone. When roosting the wings are wrapped around the body. (See picture, page 34).

'Microchiroptera' means 'small hand-winged'. They are a diverse group of bats, although all can echo-locate (many have good eyesight as well). Sinuses in the skull magnify high-pitched noises emanating from the larynx which is much bigger than it is in flying-foxes. Some Microchiropterans have elaborate facial skin decorations to focus these noises which are emitted through either the nose or the mouth. Shrews are the only other mammals which can echo-locate.

All Microchiropterans have only one claw on the wing digits, and wings are usually folded against the body while roosting (see page 34). They are mainly insectivorous although some are carnivorous or herbivorous.

Four of the five digits on the forefeet are linked by a membrane which extends from the shoulder to the toes and ankle. The sternum is shaped like a keel (although nowhere near as massive as that found in birds), and the large muscles which power flight are attached to this. The wing is a modified hand, quite similar to that found in primates. Its membranes are tougher than rubber gloves. They use wings not only for flight but to trap insects and even to swim, should they happen to fall into water (they swim quite well, although not at all gracefully).

Bats' hindlimbs will not support their weight - the femur (thighbone) has rotated 180° and the knee points backwards. The feet tendons close the toes when the bat is hanging and some Microchiropterans actually have adhesive pads or suckers on their toes to make hanging around even more secure. Bats are relatively long-lived, one Microchiropteran surviving for thirty years.

The inclusion of Megachiropera and Microchiroptera in the one order suggests that they had a common ancestor yet this may not be the case. It is possible that insectivorous bats evolved from a shrew-like ancestor. Flying-foxes on the other hand share a number of features with primates eg parts of the brain, skeletal and reproductive systems (they have erectile tissue in the head of the penis just like primates).

Bats possibly evolved perhaps 70-100 million years ago. The oldest fossils discovered in Australia are dated from about 55 million years ago. This suggests that some bats first entered Australia while it was in mid-ocean (after it separated from Gondwana and before it hit the tectonic plate on which the Indonesian islands are situatated). While Microchiropterans are known from both the Americas and Europe, and we have some families in common, there are no flying-foxes in North and South America and only fossil remains in Europe.

Reproductive equipment and behaviour

Some male bats have erectile issue in the foreskin, which enables it to 'lock' into the female. Others have a bone, the bacculum, in the penis which helps to stiffen it. Toth of these features, are also found in some carnivores, such as dogs, and some rodents. The 'locking mechanism' is found in bats that hibernate, and acts to help the torpid male complete his business. Female bats have a clitoris.

Males may have an almost permanent erection in the breeding season. Mating occurs from behind, sometimes several times in an hour.

In the flying-foxes only one ovary ovulates during a breeding season. However Michrochiropterans vary. In some only the right ovary is functional while both function in other species. Embryonic diapause occurs in some bats (as it does in kangaroos) in which the fertilized egg, or blastocyst, becomes dormant. Gestation can last from five weeks in small bats, to eight months in some flying-foxes. Birth may be headfirst or breech and there is usually only one baby, although some bats have multiple births.

*Echolocation is the finding of objects, including prey, by determining the time taken for the bat's call, usually a click, to return.

Some Microchiropterans have a tail membrane which acts as cradle to catch the baby. Others use their wings to the same end. Mother may hang head down, or pull herself horizontal to a branch for the birth (obviously gravity does not assist in such a birth). The mammary glands are in the armpit as in primates and dugongs. The baby bats may be carried around at first. and then later left in a nursery. They have milk teeth with hooks on them that allow them to cling to the mother.

FLYING-FOXES/FRUIT BATS
Suborder: Megachiroptera

LITTLE RED FLYING-FOX/FRUIT BAT *(Pteropus scapulatus)*

Description: A reddish flying-fox with light yellow or brown fur across the shoulders. Wings look transparent in flight. Large rat size. 20 cm in length with a 1 m wingspan. It lacks a tail. Males are larger than females. Common. (See picture page 34).

Habits: This flying-fox feeds on nectar, soft fruit, leaves and some pollen. It forms roosts, often with Black Flying-fox, of up to several thousand which are guarded by old males. At sunset scouts seek out food and return to communicate their find. They drink by flying onto surface water to soak the chest fur and then licking it.

Females separate from adult males after mating, forming their own camps where the single young is born. Babies emerge head first and are carried by the mother until fully furred, when they are then left in camp. The young can fly at about two months and are sexually mature at about 18 months.

Where found: Found in coastal and sub-coastal areas of much of Australia and in New Guinea. It is the most widespread flying-fox in Australia. Top End records from Fogg Dam and Holmes Jungle in the Darwin area. Also every major national park, and Cutta Cutta Caves, Manton Dam and Umbrawarra Gorge Reserves and Pine Creek/ Katherine and Borroloola regions.

Notes: I once saw this species caught and eaten by White-bellied Sea-eagle which chased them out of trees during the day. Both species of flying-foxes are preyed upon eagles, pythons and the large Rufous Owl, *Ninox rufa.* Another time I saw a lone flying-fox harassed by immature Silver-backed Butcherbirds which hopped around pecking at it. The bat responded by flapping its wings and screeching, eventually scaring the birds away.

Gunwinggu people call all big bats Gulaban, and Little Red Flying-fox Ngagayalak. They say that flying-foxes are 'good tucker'. The meat tastes rather sweet. The wings are cut off and they are cooked whole in the ashes. Everything is eaten but the gall bladder.

Flying-foxes find their way around using sight. On dark nights I have seen them blunder noisily into treetrunks, suggesting that sight does not perfectly compensate for the lack of echo-locating equipment. Like primates, echidnas and dolphins, flying-foxes have a large neocortex (see Echidna).

BLACK FLYING-FOX/FRUIT-BAT *(Pteropus alecto)*

Description: A black bat with a reddish collar, a little bigger than Little Red Flying-fox. In flight the wings appear opaque. A bat in a mango tree in Darwin is probably this species. Common.

Habits: Similar to Little Red Flying-fox.

Where found: Found around much of the sub-coastal and coastal regions of northern Australia. There is a roost in mangroves near the Sanderson Sewerage Ponds, Darwin. Records from Fogg Dam, Howard Springs, Manton Dam and Umbrawarra Gorge Reserves, and every major national park. Colonies on Mary River may have shifted because of disturbance by tour operators and others. Noted at Edith Falls in wet season feeding on paperbark blossom. Big colonies in the palms *Livistona rigida* at Mataranka Springs.

Notes: An injured Black Flying-fox, name of Dreadnought, lived in our wardrobe for some months. When fed he would reach out gently to take my hand and the teaspoon to his mouth. This species often gets caught in the barbed wire fencing around the Leanyer sewerage ponds. Gunwinggu people call this species Muro.

Megachiropteran bat roosting

Little Red Flying-fox

Ghost Bat

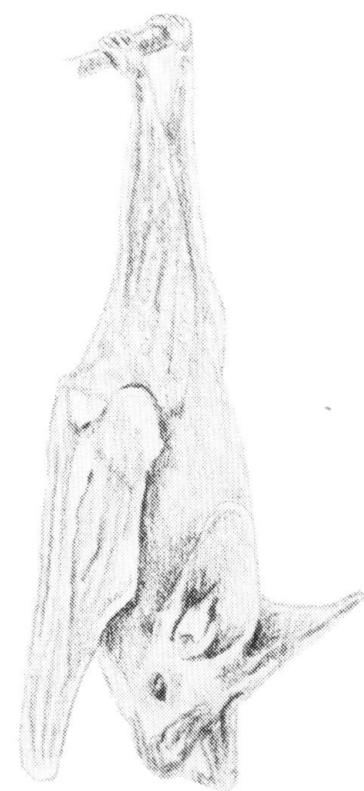

Microchiropteran bat roosting

INSECTIVOROUS BATS
Suborder: Microchiroptera

Seventy percent of all bat species are Microchiropterans. There are about 35 species of small insectivorous bats in the Top End. Most are difficult for the lay person to identify and have not been included in this book. For more information read the ***Field Guide to Bats of the Northern Territory,*** published by the Northern Territory Conservation Commission.

GHOST BAT *(Macroderma gigas)*

Description: A medium-sized, cave-dwelling, carnivorous, grey bat, with huge ears that are joined together at the base. It has big eyes (for a Microchiropteran), a noseleaf and a tragus (the long pointy flap at the base of the ear shown in the drawing, previous page) and looks rather like a demented, deformed rabbit. Rat-size 10-12 cm. Uncommon in Top End except in areas where there are big roosts. Probably rare elsewhere in northern Australia.

Habits: Ghost bats eat invertebrates and small vertebrates such as frogs and small mammals, including other bats, and birds. I discovered the remains of a carcase of a Kakadu Dunnart in one roost and Friarbird remains in another. Gouldian Finch remains have also been found in roosts.

Ghost Bats vacate roosts just after sunset. They find a vantage point from which to seek prey and then fly out to tackle it, using both sight and echo location. They drop on the unfortunate victim and kill it with bites to the head and neck, then take it back to the roost to be consumed.

The females congregate in nursing colonies, sometimes consisting of several hundred individuals. A single young is born around September-October. Like flying-foxes they carry their young for a number of weeks and then leave them behind in the nursery. These animals have a rather quiet disposition that flies in the face of their rather fierce features. They are easily disturbed, flying away from intruders into daylight.

Where found: Ghost Bats are sparsely distributed across northern Australia. Top End records from all major national parks, and Cutta Cutta Caves and Umbrawarra Gorge Reserves. They are found throughout the Pine Creek-Katherine region in suitable habitat and around Borroloola. I have found single males in the Pine Creek region with large deposits of faeces beneath their roost indicating long residence.

Notes: Fossil deposits indicate that Ghost Bats were once distributed throughout southern Australia but disappeared before white settlement. White settlement has not helped much either. Recently the site of a large Ghost Bat colony, Mt Etna, was dynamited in Queensland, a very sad event for such a beautiful and interesting animal and for those who care about them. Underground mining has partly compensated in the Top End by providing tunnels suitable for colonisation by Ghost Bats.

Ghost Bats are the largest Microchiropterans in Australia and our only carnivorous bat. It is the only Australian member of the family Megadermatidae which occurs mostly in Africa and Asia. Gunwinggu people call them Buma buma. They are not eaten.

ORANGE HORSESHOE-BAT *(Rhinonicteris aurantius)*

Description: A tiny broad-winged orange bat that lives in warm caves. This beautiful little creature has an elaborate noseleaf shaped like a horseshoe. Immature horseshoe-bats are brown. Smaller than mouse-size (5 cm). Rare. (See picture, page 36).

Habits: These bats can only live in warm humid caves at 29-32 ^{o} Celsius, as unlike other bats they do not huddle close together to preserve body heat. Little is known about breeding ecology except that they leave their caves seasonally for this purpose. They have a single young. Males of related species do not mature until a year old, females until two or three years old, suggesting that these bats may have quite a long lifespan for a small mammal.

Where found: Orange Horseshoe-bats are sparsely distributed across Northern Australia. Records from Cutta Cutta Caves and every major Top End national park. They inhabit sandstone caves in

Litchfield Park and limestone in the Pine Creek/Katherine region. Occasionally dead Orange Horseshoe-bats are found in the grills of cars travelling in Kakadu National Park during the wet season.

Note: Orange Horseshoe-bats are the only member of an endemic Australian genus *Rhinonycteris,* fossils of which have been found to be 3-5 million years old.

Orange Horseshoe-bats and other species, are susceptible to disturbance which can cause them to vacate a cave. Both locals and tourists have disrupted them within Litchfield National Park and near Katherine by climbing into caves, flashing torches and yelling.

There are four species of horseshoe-bats in the Top End (six in north Queensland).

Gunwinggu people call little cave-dwelling bats Malanbibi.

Other Microchiroptera: Northern Mastiff-bat *(Chaerophon jobensis)* This tree-dwelling species is dark-brown and grows to about 5 cm. It is found in monsoon forest and mangroves and occasionally within caves. Mastiff-bats often have heavily wrinkled lips like a miniature bull-dog. Some species have ears joined together above the forehead, others lack this feature. Some are quite at home on the ground where they scuttle around like large cockroaches. Northern Mastiff-bat is found across northern Australia and in New Guinea. Top End records from Fogg Dam, Kakadu, Keep River and Nitmiluk National Parks and Borroloola and Pine Creek regions.

Common Sheath-tailed Bat *(Taphozous georgianus)* A light-brown cave-dwelling bat about 7 cm. Sheath-tailed bats have a rather pretty face - they look like miniature dogs. Their tails are enclosed in a sheath which gives the hindlegs more mobility. Records of Common Sheath-tailed Bat from every major national park, and Umbrawarra Gorge Reserve, and Douglas River, Fountainhead/Pine Creek/Katherine and Borroloola regions.

Different types of Top End Microchiropteran bats

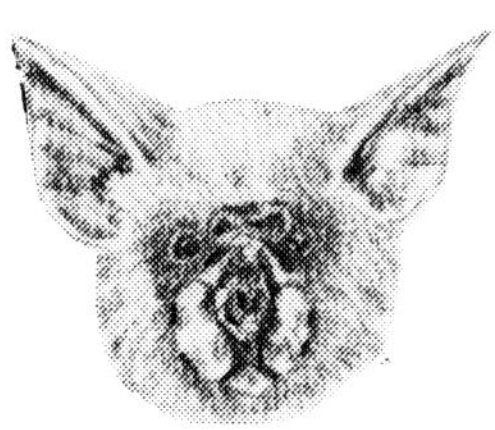

Orange Horeshoe-bat

Mastiff-bat

Sheath-tailed Bat

RATS AND MICE
Order Rodentia

Rodents have one pair each of upper and lower incisors. Hard enamel on the front surface of the incisors resists wear while softer dentine at the back wears more rapidly thus ensuring a chisel-like edge. These teeth grow throughout life and must be continually worn down.

All rodents in Australia belong to the one family, Muridae. Some Australian groups, *Conilurus, Mesembriomys, Notomys, Pseudomys* and *Zyzomys* ,entered Australia and New Guinea 10-15 million years, and are called the 'old endemics'. Others, the 'new endemics' such as *Rattus* species, came later, perhaps only a million or two years ago. Still other species have been introduced since European settlement.

The old endemics are relatively long-lived, and most have few nipples and produce few young, although there are exceptions to this rule. The 'new endemics', have many nipples and produce many young.

The ranges of many native rodents have shrunk, some before European settlement in the 1770's, possibly as a result of decreasing rainfall. Since then ranges have contracted because of habitat loss caused by sheep, cattle and rabbits. Cats and foxes, introduced by settlers, became efficient predators causing some species to become extinct.

WATER-RAT *(Hydromys chrysogaster)*

Description: A dark brown, large, solitary rat found near or in water. Rabbit-size. 30 cm. (excluding tail). It is blunt-nosed with tiny ears and eyes and has soft, thick, water-repellant fur (elsewhere in Australia its fur may be light-coloured). Its tail is furred with a white tip. The hindfeet are partially webbed. Relatively common. (picture next page).

Habits: Unlike most rodents, Water-rats are carnivorous, feeding on aquatic animals, eg fish and mussels, and on carrion and occasionally, waterbirds. They make a burrow along the water's edge. Water-rats do not breed until they are a year old, having up to four young a litter. I have seen most animals in early morning in the tropics.

Where found: near or in water. Found in most Australian waterways, and in New Guinea. Top End records from Kakadu, Keep River and Litchfield National Parks, and Fogg Dam and Umbrawarra Gorge Reserves, Marrakai Plains and the Pine Creek and Borroloola regions.

Notes: They were once trapped for their beautiful fur. Fortunately for such a slow-breeding animal this was discontinued. Gunwinggu people call this animal Yirrku.

Similar species: Black-footed Tree-rat *(Mesembriomys gouldii)* is of a similar size and colour and its tail also has an extensive white tip. However its coat is coarser, and it is more slightly built than Water-rat. Black-footed Tree-rat inhabits trees in open forest but is occasionally found on the ground. It is nocturnal and solitary, with a rather unfriendly disposition. One to three young are produced and are dragged behind the mother. The females become sexually mature at about three months. They are prodigious leapers. One, disturbed beside the Kakadu Highway, leapt six metres up a bank beside the road to escape. Rare to common. (picture, previous page).

Black-footed Tree-rat is found in open forest across coastal and sub-coastal Northern Australia and on Melville and Bathurst Islands. Records from Gurig, Kakadu and Litchfield National Parks and also at Manton Dam Reserve and Mary River.

Golden-backed Tree-rat *(Mesembriomys macrurus)* is smaller than either of the above species and is grey above with an orange dorsal stripe. It has a long tail with a brush. This species inhabits a range of habitats from coastal dunes to eucalypt woodland and is known from the northern Top End and northwestern Western Australia. Rare in most areas.

Brush-tailed Rabbit-rat *(Conilurus penicillatus)* is about 15-20 cm (excluding tail). It is a similar colour to the Black-footed Tree-rat but looks quite different, more like a long-tailed rabbit. It has a rounded head with bulbous eyes and large rabbit-like ears. It has a brush on the end of its tail which can be black or white. Rabbit-rat is also arboreal . It has two young. *'Conilurus'* means 'tailed rabbit' in Greek.

It is found in open forest/woodland, and in coastal areas across the Top End and into northern Western Australia. Recorded from Gurig and Kakadu National Parks. Also Melville and Bathurst Islands and near Borroloola. Possible sightings from Douglas River and near Mount Wells. Widespread but sightings are uncommon. (picture, previous page).

GRASSLAND MELOMYS *(Melomys burtoni)*

Description: A grey or yellow/brown nocturnal rat with a scaly, partly prehensile tail which is often darker above. It has soft fur, round, naked ears and broad hindfeet. 14 cm (excluding tail). Common in some areas. This rat is one of the mosaic-tailed rats, so-called because the scales on the tail do not overlap as they do in most other rats. (picture next page).

Habits: This rodent is partly arboreal and makes a nest, often in pandanus in which it spends the day. It is omnivorous. Females have up to three young which mature in three weeks.

Where found: In monsoon forest and wetter grassland across northern Australia and down the east coast. Top End records from Fogg Dam, Holmes Jungle, Howard Springs and Manton Dam Reserves, every major Top End national park and the Pine Creek/ Douglas River and Borroloola areas. This genus is mainly found in New Guinea.

Notes: Considered a pest in Queensland sugar-cane fields where it is known to damage cane. It is one of several medium-sized rodents called Mulbu by Gunwinggu people.

Similar species: Can be mistaken for some other rodents and for Fawn Antechinus. Tail is unmistakeable if seen in a good light and fur lacks spines found in most other rodents of similar size and habitat. Ears are shorter and rounder than those of Fawn Antechinus and snout is rounded not pointed.

DELICATE MOUSE *(Pseudomys delicatulus)*

Description: A tiny nocturnal sandy-coloured mouse found in woodland. It is clearly demarcated white underneath. The scaly tail is as long as its body and it has big eyes and big round ears. The tiniest Australian rodent, it measures about 6 cm in length (excluding tail). Relatively common in some areas.

Habits: This mouse constructs a burrow in sandy soil in which it sleeps during the day. It has up to four young and can raise several litters in good seasons.

Where found: It occurs in eucalypt forest and woodland, where there are sandy soils, and is found across Northern Australia and down the east coast. Top End records from every major national park, and from Douglas River and Borroloola regions. I found a Delicate Mouse rummaging around quite unconcernedly in the cold ashes of a barbecue at Burdalba camp site in Kakadu National Park.

Notes: *'Pseudomys'* means 'false mouse'. This is the largest genus of native Australian mice. Females have only two pairs of teats. Gunwinggu people call this mouse Kijbuk.

Similar species: no other rodent is as small as Delicate Mouse. **Forrest's Mouse** ***(Leggadina forresti)*** is plump, unlike the slender Delicate Mouse and has a tail much shorter than its body length of 8-10 cm. It is a beautiful snub-nosed little animal with grey fur and small round ears which has up to four young and several litters a year when conditions are favourable. It inhabits grassy plains and forest along creeks and is found in northwestern and central Australia. Top End records from Cutta Cutta Caves, Gregory and Kakadu National Parks and from a pitfall trap near Edith Falls during the Mt Todd survey. Also found near Borroloola. Uncommon.

House Mouse ***(Mus musculus)*** grows to 9 cm, and has five pairs of nipples . I have never seen a house mouse with the neat appearance and colouration of Delicate Mouse. Because House Mouse can have much bigger litters they pose a threat. to native mice. Recorded from Darwin, Howard Springs and Kakadu National Park.

Western Chestnut Mouse ***(Pseudomys nanus)*** is still larger, growing to over 12 cm. It is grizzled orange/brown above and lighter underneath with a pronounced light-coloured eye-ring and short round ears. Its scaly tail is shorter than the body length and is dark above and light below. Call is a high-pitched whistle. It is found in dense grassland, often near watercourses across the Top End and into northwest Western Australia. Records from Gregory, Kakadu, Keep River and Nitmiluk National Parks and the Mary River, Pine Creek/Katherine and Borroloola regions. Relatively common in suitable habitat.

Grassland Melomys

Delicate Mouse

Forrest's Mouse

Western Chestnut Mouse

DUSKY RAT *(Rattus colletti)*

Description: A dark brown, small-eared rat found in or near wetlands. It has coarse, spiny fur which is paler underneath, often yellowish. The tail is shorter than the body length. 15-20 cm (excluding tail). Abundant within a limited area.

Habits: The Dusky Rat inhabits Top End wetlands and feeds on the corms of sedges, and seeds. During the dry season they inhabit burrows, or cracks in the soil which are also home to unsociable neighbours such as Death Adders. However they also offer shelter from grassfires that rage through the wetlands every dry season. During the wet season the Dusky Rat migrates to higher ground as the wetlands flood. These rats are considered a pest by rice-growers because they burrow into causeways around paddie-fields. Breeding starts at the beginning of the wet season and continues into the early dry. The females can breed when they are a month old and have large litters of up to twelve.

Where found: This rat is found only in the Top End. Records from Fogg Dam and Holmes Jungle Reserves, and Kakadu and Litchfield National Parks and Douglas River and Mt Bundey areas.

Notes: *Rattus* species have a scaly tail and most have spines in their fur which is often coarse to the touch. They are relatively difficult to tell apart.

Gunwinngu people eat this rat. They call it, and those mentioned below, Mulbu.

Similar species: Black Rat/Plague Rat ***Rattus rattus.*** This introduced animal is found in similar areas to Dusky Rat. It is not necessarily black - pale and mid-brown specimens are common. However the coat is smooth and sleek, unlike that of Dusky rat. It also lacks Dusky Rat's yellowish flanks and underside, and its tail is usually much longer. The ears are big and paper-thin. Black Rat is known to eat small vertebrates as well as insects and vegetable matter. It matures at a much slower rate than Dusky Rat.

Having kept Black Rats found in the wild as pets, I have found them docile and trusting, curling up in one's hand to go to sleep. This rat was introduced to Europe by the Crusaders and arrived in Australia with the First Fleet. Unfortunately its fleas carried bubonic plague. Found around Darwin and into Kakadu National Park.

Pale Field-rat ***(Rattus tunneyi)*** is paler than Dusky Rat, but also has a short tail. It has beautiful toffee-coloured or grey-brown lustrous fur, paler underneath, protruding eyes and short round ears. It has a pleasant disposition, much like that of the Black Rat. Inhabits grassland near watercourses and open forest. Pale Field-rats can be a minor pest. They destroyed the crops of settlers at Port Essington on Cobourg Peninsula in the 1840s.

They breed during the dry season having litters of up to seven, maturing at about a month. Pale Field-rat is found around the coastal and subcoastal regions of northern and eastern Australia. Records in the Darwin area from Fogg Dam and Holmes Jungle, where Dusky Rats are also found. Also Gregory, Gurig, Kakadu and Litchfield National Parks and the Borroloola region. It is found as far inland as the Pine Creek/Katherine region.

COMMON ROCK-RAT ***(Zyzomys argurus)***

Description: A light-brown rat sometimes seen nosing around rocky outcrops at night. It is distinctly roman-nosed with big eyes and round ears. The hindfeet are slender and the tail is thick at the base. 10-12 cm (excluding tail). Common in suitable habitat.

Habits: The quiet observer can sometimes see Common Rock-rats at night by sitting still and watching. They come out and snuffle around quite oblivious to the onlooker. They eat vegetable matter and insects (also rice pudding and custard as I discovered recently in Barramundi Gorge, Kakadu). The females become sexually mature at five months old, having 2-4 young a litter.

Where found: Rocky outcrops across most of northern Australia. Top End records from Cutta Cutta Caves, Manton Dam and Umbrawarra Gorge Reserves and all the major national parks. Also records from Douglas River/ Pine Creek/ Katherine and Borroloola regions.

Note: Rock-rats drop their tails easily when captured. This is possibly a defence mechanism such as exists in some lizards.

Similar species: Northern Hopping Mouse *(Notomys aquilo)* A little brown nocturnal mouse with white underparts, long hindlegs, a hopping gait, and a very long tufted tail. It lives in burrows in sandstone countryand also in coastal sandhills. Mouse size 9-11 cm (excluding tail) . Up to 3-4 are born at a time and they are furred, unlike the young of rats. Records from Arnhemland, Roper River and Groote Eylandt. It is possibly found in Kakadu National Park. Notify NT Conservation Commission and/ or Park authorities of any sightings.

Australian Hopping-mice look very similar to North American hopping-mice and the jerboas of Africa and Asia, but are not closely related.

Large Rock-rat *(Zyzomys woodwardi)* is a larger rat, reaching 17 cm. It is grey-brown in colour and is much rarer than Common Rock-rat. It is found in Kakadu National Park and Arnhemland.

OTHER NATIVE MAMMALS

This section covers other mammals, native to the Top End.

Order: Carnivora
Family: Canidae

DINGO *(Canis familiaris dingo)*

Description: A small to medium-sized dog (0.8-1 m.), usually fawn/yellow with white points. Occasional dark specimens noted in the Top End. Common.

Habits: Dingoes are usually solitary animals although a small group may hunt together to pull down larger prey. They are nocturnal but are often seen during the day. They take small prey such as mammals and birds and will also eat carrion. Dingoes have been long blamed for stock losses, especially among sheep in southern and western Australia. The world's longest fence was erected across Australia to keep dingos out of sheep country to the south and east. They are poisoned with 1080 and notices announcing the setting of such baits often appear in the *Northern Territory News*. They do not breed as often as domestic dogs, only once a year. Younger non-breeding animals may help new parents with young.

Where found: Dingoes are found all over Australia, except Tasmania, In the Top End it frequents savannah woodland, and open forest and can often be seen on roadsides. I have seen them in Darwin at the Casuarina Coastal Reserve, Holmes Jungle and around Sanderson sewerage ponds. Records also from Fogg Dam, Howard Springs, Manton Dam and Umbrawarra Gorge Reserves, and every major national park, and in the Mary River,, Fountainhead/ Pine Creek and Borroloola regions.

Notes: Dingoes were brought to Australia by Aboriginal people possibly as recently as 3000 years ago and appear to be indistinguishable from the jackal found through Asia and India. Dingoes have been implicated in a number of attacks on children, the most famous being the Chamberlain case at Ayers Rock in 1980. Gunwinggu people say they have been known to steal their babies.

Beware of half-tame dingoes in parks and reserves. Gunwinggu people call this animal Dulkun.

Similar species: Very similar to many varieties of domesticated dogs.

Thylacine ***(Thylacine cynocephalus)***, or Tasmanian Tiger, is considered extinct. This was a dog-like, carnivorous marsupial with a striped back, long tail and sloping hindquarters, growing to about 130 cm. It had a backward-opening pouch. The Thylacine was found all over Australia and New Guinea. Its numbers then declined, possibly because of competition with the dingo. However dingoes were not present in Tasmania and Thylacines were relatively protected there until European settlement. They soon became known as sheep-killers and the Tasmanian government introduced a bounty in 1888. Thylacines rapidly declined (disease might have been a factor as well) and the last known survivor died in 1936.

There are a number of Aboriginal rock-paintings of Thylacines in the Top End, the best known at Ubirr in Kakadu. According to Arthur Palmer, ethnographic consultant, Aboriginal people in both the Victoria Rivers district and on the east coast of Arnhemland, insist that the species still exists. They can describe the 'rear pouch, teats, genitalia' and can name localities where it lives and has been recently sighted. Mr. Palmer states that at 'a land claim hearing in 1981, oral evidence was given of 'lion or tiger dreaming' to support the claim book submission on this area.' The Tasmanian race was sandy in colour. Top End Thylacines may well be darker.

Notes: Thylacine tracks would differ from those of a dog/dingo. Thylacines have five toes on the forefeet, dogs have four. There are stuffed Thylacines in the Adelaide and other southern museums.

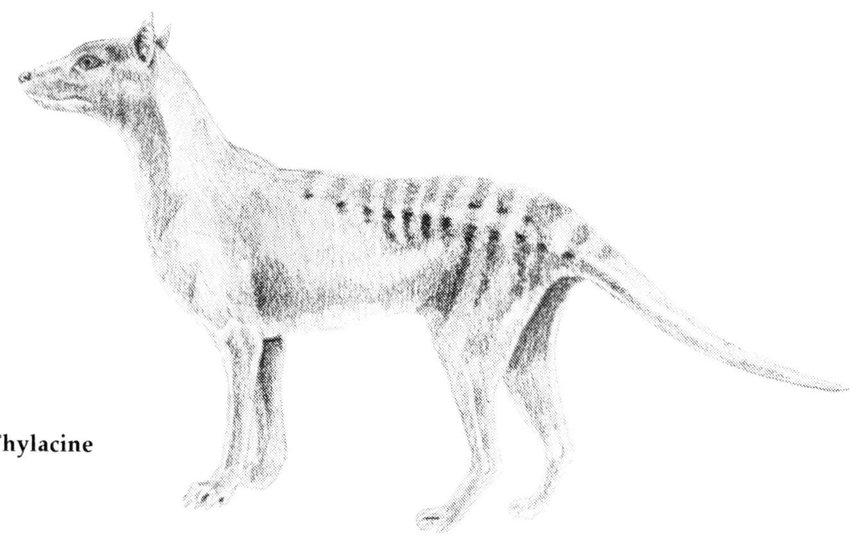

Thylacine

Order: Sirenia
Family: Dugongidae

DUGONG ***(Dugong dugon)***

Description: A large, grey, herbivorous, marine mammal that grows to about 2.5 -3 m. and weighs up to 400 kg. It has a horizontal tail fluke like a whale, small eyes and a bulbous downward-pointing bristle-covered snout. Males have a pair of upper incisors that protrude from below the upper lip. The only other outward indication of sex is that the distance between the anal and genital openings of males is farther apart than that of females. Relatively common in suitable habitat, but vulnerable. (picture next page).

Habits: These harmless mammals feed on beds of seagrass that are present off the Top End coast. A couple on a yacht moored in Fannie Bay, Darwin used to feed resident dugongs lettuce leaves. They feed both day and night and will enter shallow water for this purpose. They see and hear quite well and are vocal, giving out a range of calls. They become sexually mature at 9-15 years and live to about 70 years old. Mating occurs with the male rolling the female on her back with his tusks. A female will turn her back on the male when she is not ready, but whether this puts him in his place or not I don't know.

There is a gestation period of about one year. Females give birth in shallow water where they are safer from sharks and crocodiles. Young are carried on the mother's back unless they are feeding.

Pregnancy occurs every 3-7 years and lasts about 13 months when the female bears a single young. Nipples are found at the base of the pectoral flippers.

Where found: Possibly 90% of the world's dugong live in Australian waters. They can be found all around the coast of northern Australia especially where there are large beds of seagrass. In Darwin they can sometimes be seen from East Point and from Casuarina Coastal Reserve. There are also records from Fogg Bay west of Darwin, at Gurig National Park and near Borroloola.

Notes: Dugongs are preyed upon by saltwater crocodiles and sharks. They are also killed by Aboriginal people for food. The flesh is delicious, tasting like pork. They were hunted by Europeans for food and for oil until the late 1960's when they were finally protected. Gunwinggu people call the dugong Mandingnganj.

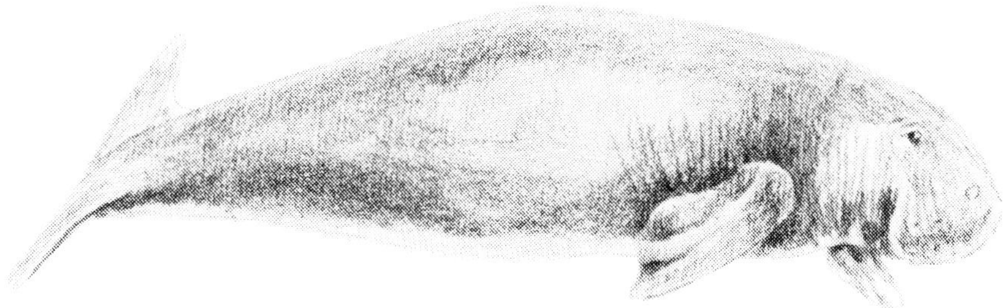

Dugong

BOTTLE-NOSED DOLPHIN *(Tursiops truncatus)*

Description: A plain grey, beaked dolphin with a small curved dorsal fin. Its mouth contains 80-100 undifferentiated peg-like teeth and it has a small ear-opening behind the eye. 3 m. Common.

Habits: Like other dolphins, this species has a large vocabulary of high-pitched sounds. They are often gregarious and may swim close to boats or the shore. It can reach speeds of 35 km/h. Brian Whitehead and I saw one at Fog Bay chasing a fish. The dolphin skidded into an acute turn at great speed, like a racing car at the Grand Prix.

Males become sexually mature at between ten and twelve years, females much earlier. Dolphins give birth after a gestation period of a year. Birth is breech and is attended by other dolphins which assist the baby to the surface to breathe. The single youngster is about one metre long at birth and suckles for over a year. Other dolphins will care for the baby while the mother is feeding. Dolphins can live to be 35 years old.

Where found: Recorded around Australia. In the Top End I have noted several in Frances Bay, Fog Bay, Coburg Peninsula and alsoin the waters around Nhulunbuy.

Notes: Taiwanese fishermen have drowned thousands of dolphins in fishing nets off the north Australian coast. In the 1980's I worked with a researcher who was attempting to modify nets so that dolphins would recognise them and stay away. As part of the job we had to clean the skeletons. I rendered down over a hundred dolphins, mostly on the kitchen stove. Unfortunately he died before the work was complete, but I believe that such nets have now been successfully tested.

Dolphins have proportionately smaller lungs than other mammals but can exchange more oxygen at each breath than most. The dorsal fin contains no bones - just muscle. A dolphin's penis is anchored to a small bone, a remnant of the dolphin's pelvis and femur, and is curled up in a pouch made of the foreskin unless erect. The penis contains special tissue that ensures that it is always semi-erect.

REPTILES
Class Reptilia

There are three extant groups of reptiles in Australia, crocodiles, turtles, and snakes and lizards. They are all covered in this section as well as one group that is probably extinct, the Plesiosaurs and Ichthyosaurs.

Reptiles like amphibians are ectothermic, that is they cannot regulate their body heat by internal means. However, unlike amphibians they have an impermeable skin.

Some reptiles have one penis, others have two hemipenes. 'Hemipene' literally means 'half penis'. These latter organs lie inside the body and are everted for mating. Reptiles have a similar reproductive arrangement to monotremes with all ducts opening into a sinus with one common opening, the cloaca. Some lay eggs and others give birth to live young.

PLESIOSAURS AND ICHTHYOSAURS
Subclass: Euryapsida

These animals are presumed extinct. However as some locals are convinced there is a Loch Ness look-alike lurking in the vicinity of Bynoe Harbour, and because there are fossil remains of both animals in Darwin Harbour, I have included a brief note on them.

'Plesiosaur' means 'near reptile' and 'Ichthyosaur' means 'fish reptile'. The plesiosaur had a very long neck and four flippers and some species grew to 13 m in length. One theory is that the Loch Ness monster is a plesiosaur. The ichthyosaur had a long beak, four flippers and a spine that curved along the base of the lower lobe of the tail. Ichthyosaurs produced live young which have been found, fossilised inside the mother. Both of these animals ate fish. This should reassure those boaters venturing into Darwin waters.

Read Dr. Peter Murray's paper (1987) on fossil discoveries in Darwin Harbour, given in the reference section of this book.

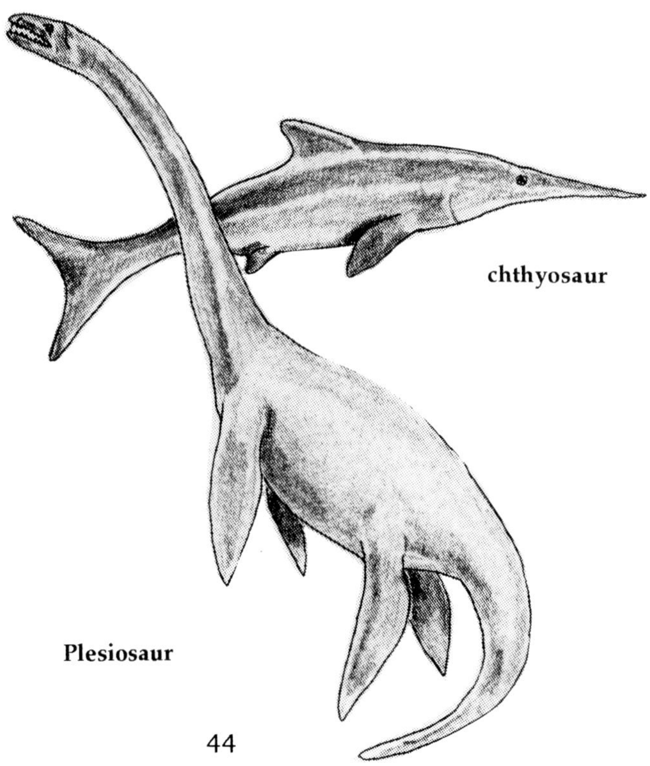

CROCODILES
Subclass Archosauria:
Order Crocodilia

Crocodiles appeared about 200 million years ago (the same time as mammals). They descended from the subclass Archosauria or 'ruling reptiles', as did birds. There are about 27 extant species of crocodiles today, two of which are found in Australia. 'Crocodile' comes from Greek 'krokodeilos', a name for a small lizard which the ancient Greeks thought resembled the Nile Crocodile.

Unlike other reptiles, crocodiles have a four-chambered heart and a secondary palate. They also have a few skeletal extras, eg bones in their eyelids and osteoderms (see Skeletons). They have an extremely efficient circulatory system. When the crocodile dives a valve between the left and right ventricles closes, restricting flow to the lungs where it is not needed and reserving oxygenated blood for the all important brain and heart.

Crocodiles have internal nostrils that can be closed with an external valve. Other valves within the throat close the mouth from the airpassages thus enabling the crocodile to hold prey underwater. The ears which are of a similar structure to those of birds, can be closed off by earflaps. A third eyelid, the nictitating membrane, protects the eyes underwater. The eyes have a tapetum lucidum, like many other vertebrates. This causes their eyes to shine pink or red by torchlight at night.

The crocodile has only one penis, unlike snakes and lizards which have two hemipenes, and the female crocodile has a clitoris! The penis is solid unlike the hemipenes of snakes and lizards which are hollow. To mate the male climbs onto the female's back and positions his tail under hers. Apparently it is quite acceptable among crocodiles for the female to initiate courtship.

Crocodiles either build a nest of vegetation or dig a hole in which to lay their hard-shelled eggs . Unlike mammals and birds whose sex is determined at gestation, the sex of crocodiles (and turtles) is determined by the temperature of the egg-chamber within the nest. Crocodile can deliver a larger degree of parental care than any other reptile. They respond to the cries of hatching youngsters and may actually assist some to hatch by squeezing the eggs gently between their teeth. They then transport the babies to the water, by holding them gently in their mouths. Adult crocodiles vocalise with grunts and bellows.

Crocodiles can go without food for long periods. Indeed freshwater crocodiles seem to go most of the dry season without food.

Crocodiles were hunted heavily until the 1960s. The freshwater crocodile became protected by legislation in 1964 and the estaurine in 1971.

JOHNSTON'S/FRESHWATER CROCODILE *(Crocodylus johnstoni)*

Description: A rather shy crocodile with a long slender snout. Grey with variable darker mottling and cream below. There is a single row of enlarged nuchal shields (scales found on the nape). See illustration next page. Grows to 3 m, but is usually much smaller. Common in suitable habitat.

Habits: Eats fish, crustaceans, waterbirds. Lays eggs in sandbanks along rivers. Many are eaten by goannas and pigs and nests may be trampled by stock as well. This species can sometimes be seen sunning itself on rocks. When disturbed on land this crocodile can launch itself right into the air in an attempt to get back into deep water. Like other species of crocodiles it demonstrates a homing instinct if removed from its territory.

Where found: Found in freshwater throughout northern Australia and the Top End. They have been recorded from small billabongs and creeks in and near Darwin, and in most rivers - Douglas, Daly, Edith, Flora, Katherine and Victoria Rivers and associated creeks and rivers in the Borroloola region. In Kakadu National Park Freshwater Crocodile is present at Yellow Waters, Twin and Jim Jim Falls and Barramundi Gorge. It is found in most other Top End national parks and Manton Dam and Umbrawarra Gorge Reserves.

Note: This crocodile is relatively harmless if left alone. Unfortunately people have been observed throwing stones at them and otherwise annoying them, in Kakadu National Park and elsewhere. Ask local rangers in parks and tour operators if freshwater crocodiles are present in a pool and what areas of that pool they inhabit. Then avoid those areas. This species, and Estuarine Crocodile (below) are both called Kinga by Gunwinggu people.

Freshwater Crocodile has a long, slender snout. That of the Estaurine Crocodile is relatively shorter and broader. In freshwater crocodiles the nuchal scales on the nape form a single row, a little behind the smooth head scales. In Estaurine crocodiles the nuchal scales are in two rows and are widely separated from the smooth scales of the head.

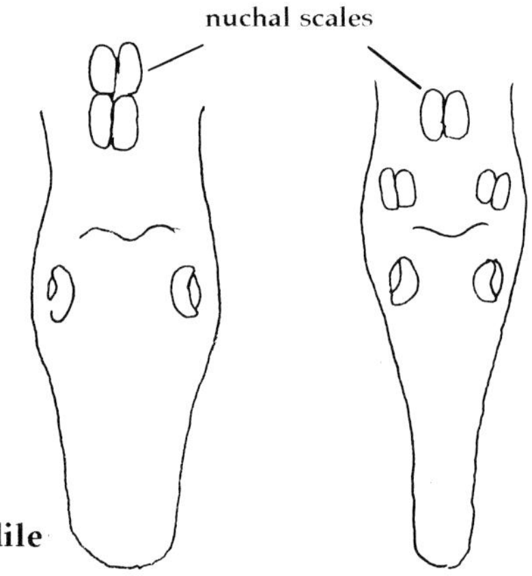

ESTAURINE/SALTWATER CROCODILE *(Crocodylus porosus)*

Description: A large crocodile with a broad head and snout. Grey/brown with dark mottling and cream/white below. There are two rows of enlarged nuchal (nape) shields and two ridges running from the eyes toward the snout. Has been recorded at 7 m. in length and weighing in at around 1000 kilograms. However most individuals are much smaller. Common in suitable habitat.

Habits: Smaller estaurine crocodiles feed on waterbirds including Magpie Geese, and fish. Larger ones will take correspondingly larger prey including wallabies and even stock. They have also have been noted leaping out of the water to take flying birds. A 3m crocodile at Sanderson Sewerage Ponds used to jump 2 m into the air to snatch kites from a railing, to the consternation of birdwatchers and sewerage pond workers. The Conservation Commission has saved the day by erecting a 'No Swimming-Crocodiles' sign at the main entrance.

Estaurine crocodiles lay eggs in nests made in vegetation, usually during the wet season. Mortality is high, and is largely caused by flooding and predation by monitors. Crocodiles guard their nests. A female at the Darwn Crocodile farm charged visitors at the fence who were too close to her nest. Eggs take 2-3 months to hatch. Estaurine crocodiles can live to around 80 years old.

Where found: This crocodile is **NOT** confined to saltwater. Estuarine crocodiles are occasionally reported from Berry Springs, and Twin Falls in Kakadu and six years ago two were caught downstream from Wangi Falls in Litchfield Park, although none have been sighted there since. They can walk overland - skeletal remains are occasionally found in the bush miles from water.

Estaurine Crocodiles are found in coastal India, southeast Asia, the Philippines and northern Australia. In the Top End they are often removed from Darwin Harbour by the Conservation Commission, 123 in 1991/92 and 155 in 1992/93 (courtesy of Stuart Wohl). They are occasionally found in Darwin itself. One was reportedly run over by a taxi on Bagot Road in Darwin some years ago. They are found in most parks, eg Fogg Dam, Shady Camp and Swim Creek and Gregory, Gurig,

Kakadu (Jim Jim and Nourlangie Creek, and Mary, East Alligator and South Alligator Rivers), Keep River and Litchfield (Finniss and Reynolds Rivers) National Parks, Roper River and in the Borroloola region.

Notes: Some traditional Aboriginal people, including my adopted Gunwinggu son, can call these crocodiles up to them and apparently get to know individuals well enough to bathe in relative safety around them. They also have ways of scaring crocodiles away, methods that even young children are taught. However even Aboriginal people are taken from time to time. My adopted son says that the presence of buffalo and horses caused numbers of big crocodiles to increase and that ow that these animals have been shot out of many areas, those crocodiles are becoming a risk.

Estaurine crocodiles are **DANGEROUS.** Be guided by Park rangers.

Some tour operators ram crocodiles with boats to 'give tourists a thrill.' In fact one operator told me that 'everyone does it'. Anyone behaving in this manner should be reported to either the Conservation Commission or the Australian Nature Conservation Agency (formerly Australian National Parks and Wildlife). Follow up with a letter to the Minister for Tourism and newspapers. As Aboriginal elders said in response to this sort of behaviour, 'better the crocodiles stay and the tourists go.' If we want a sustainable tourism industry then operators who do such things must be checked. Thanks to a far-sighted Conservation Commission, education programmes have led to increased public awareness within the Northern Territory, and at least some of us are are learning to live with crocodiles.

At Fogg Dam one morning Hilary Thompson and I noted a 2 m estaurine crocodile lying across the causeway. A flock of Little Corellas gathered nearby, on the ground, apparently considering the crocodile an unwanted intrusion. As we watched a bird darted forward with crest and wings raised and screeched at the crocodile. It then fled back to the bosom of the flock. Then another darted forward. This went on for ten minutes or so until the crocodile decided it had had enough and slid back into the water. Hilary Thompson reports seeing a Rufous-bander Honeyeater 'dive-bombing' an estaurine crocodile lying in shallow water at Fogg Dam.

To anyone who thinks that working as a field biologist is all fun, imagine standing thigh-deep in the remote parts of the Mary River (my colleague and I were dropped in by helicopter) on a pitch-black night, in pouring rain, trying to sample fish. The Mary is well-populated in some stretches by very large 'salties', as estaurine crocodiles are called locally. However my colleague assured me that at our location there was only one every kilometre or so. I was not reassured. However this night and those that followed had one positive effect. I lost my childhood fear of the dark.

TURTLES AND TORTOISES
Order Testudines

This book only deals with freshwater turtles.

Turtles, another ancient group, appeared about 200 million years ago. One of the most spectacular of these was a giant (2 m) horned turtle that occupied parts of Australia until 50 000 or so years ago. It was a representative of a group also known from South America.

The sex of adult turtles can be distinguished by tail length. Males have a long, thick tail which is curled under the female when mating. Females have a short tail. Like crocodiles, male turtles have a single penis and female turtles have a clitoris.

Turtle eggs are a favoured food of some Aboriginal people such as the Tiwi. When boiled they taste much like fowl eggs. However the white does not set and non-Aboriginal people and Aboriginal people who have lost their taste for bush tucker, are somewhat put off by them. Turtles are cooked in their shells and in their own juices. The large marine ones are baked in hot stones in a hole in the ground. Hot stones are placed inside the turtle after the entrails are removed. Freshwater turtles are just thrown into a small fire.

PIG-NOSED/PITTED SHELL TURTLE *(Carettochelys insculpta)*
Family: Carettochelydidae

Pig-nosed Turtle is the only surviving species of a family that dates back at least 40 million years. It was originally believed to be confined to New Guinea although Aboriginal people painted it on rock walls for centuries before its 'discovery' in 1970. (Picture next page).

Description: The head of this turtle has a short fleshy snout hence the name. It has flippers, like a sea-turtle, and the shell lacks scutes (see Skeletons, instead being covered with a soft pitted skin. Grows to around 0.5 -0.75m. Uncommon.

Habits: Is omnivorous, eating both fruits and crustaceans. Females lay eggs in a shallow nest made in sandy banks late in the dry season.

Where found: Found in aquatic habitats across the Top End and in New Guinea. In Kakadu they are found in Jim Jim and Nourlangie Creeks, Gunlom and Twin Falls, and the upper reaches of the South Alligator and Mary Rivers in south eastern Kakadu National Park. Gunwinggu people report them in the East Alligator River. Also found in the Daly, Douglas and Victoria Rivers.

Notes: This turtle appears to be able to absorb oxygen directly from water through networks of capillaries in its mouth. Other turtles, so I'm told, are able to absorb oxygen through the capillaries in their backsides. However turtles have lungs as well.

Gunwinggu people call this turtle Warradjan. It was traditionally caught by a couple of methods. Either it was seized as it was laying eggs, or the fruit of aquatic pandanus was thrown into the water. This attracted the turtles which, according to my Gunwinggu son, were seized as they ate.

OTHER FRESH-WATER TURTLES
Family: Chelidae

This family is found in Australia, New Guinea and South America. Members have ankle-joints and four or five-clawed, webbed feet, and withdraw their heads under the shell sideways.

In the Top End locals call all Chelidae 'turtles'. Elsewhere in Australia they are called 'tortoises'.

NORTHERN SNAKE-NECKED TURTLE *(Chelodina rugosa)*

Description: A dull-grey turtle with a shell length of about 40 cm. It has a long neck (longer than its body) and an oval shell. It is sometimes called Long-necked Turtle. Common.

Habits: this turtle may bury itself in the mud during the dry season when ephemeral waterholes dry up, or travel overland seeking more permanent water. Many die, picked off in their weakened state by White-bellied Sea-eagles and other creatures. The female lays up to about twelve eggs in mud. The young are preyed upon by fish, goannas, water-rats and snakes. In turn they feed on fish, crustacea and mussels. And so the cycle of life goes on.

Where found: Found across northern Australia. and inhabits creeks and billabongs all over in the Top End. Recorded at Rapid Creek in Darwin, and at Berry Springs, Fogg Dam and Howard Springs Reserves Also in Kakadu and Nitmiluk National Parks. Other records from Douglas River, Lambells Lagoon near Humpty Doo, Marrakai Plains, Mary River wetlands, Wildman River. and the Borroloola region.

Notes: There are five species of Long-necked Turtle in Australia. Some have inguinal glands which emit a milky fluid with a frightful smell. Southern Aboriginal people remove these glands before cooking the tortoise. However these glands seem to be lacking in Northern Snake-necked Turtle, and Gunwinguu people put them straight on the fire. Paul Horner notes that the related species*Chelodina novaeguineae,* recently recorded from the Northern Territory, has offensive glandular secretions. Aboriginal people hunt Long-necked Turtle by feeling around with bare toes in the mud of billabongs and creeks. They and other species are also caught on fishing lines. Gunwinggu people call this species Almang (with a lazy 'ng').

Pig-nosed Turtle

Long-necked Turtle

Northern Snapping Turtle

Yellow-faced Turtle

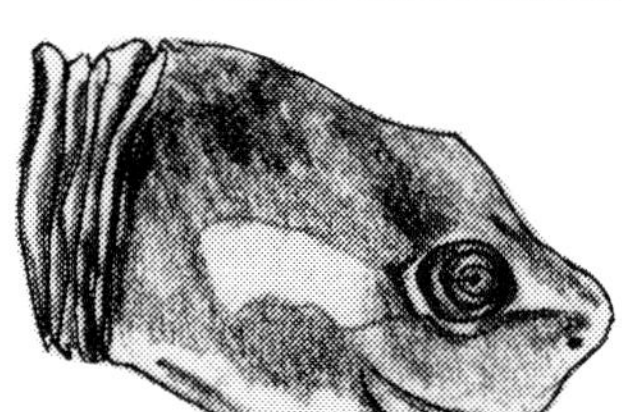

NORTHERN SNAPPING TURTLE ***(Elseya dentata)***

Description: A short-necked turtle, with a plain face and tubercules on its neck. 35 cm. Rear of shell is serrated in young specimens. It has webbed feet with five clawed digits. Common.

Habits: Omnivorous, feeding on the fruit of *Pandanus aquaticus,* white plum, *Syzygium armstrongii,* crustacea and small fish. It is not shy about giving stray fingers a nasty bite. so if you catch one on a fishing line watch out! Lays about 15-20 eggs.

Where found: Can be found in creeks and billabongs anywhere in the Top End, including around Darwin. Records from Berry and Howard Springs Reserves, and Kakadu, Litchfield and Nitmiluk National Parks.

Notes: Gunwinggu people call this species Ngarrdirrwa.

Similar species: Saw-shelled Turtle (*Elseya latisternum*) was reported from the Top End in a 1987 paper by Max King and Paul Horner of the Northern Territory Museum. Enthusiasts who wish to know more, can read the paper, given in the bibliography at the end of the book.

Yellow-faced Turtle *(Emydura victoriae)* This short-necked turtle looks similar to Northern Snapping Turtle but has a pale stripe down its face. Found throughout northern Top End and northern Western Australia it can be seen locally in Berry Springs, Howard Springs and Umbrawarra Gorge Reserves and Kakadu and Litchfield National Parks. (Picture previous page).

LIZARDS

Order: Squamata
Suborder: Sauria

Lizards differ from snakes in that most have ear openings and legs, and most (except monitors) have a broad, fleshy tongue. Snakes usually have no trace of limbs (except pythons). and all have a forked tongue. Lizards usually have tails as long, or longer than their bodies while snakes have relatively short tails. Most lizards, unlike snakes, have at least one eyelid, enabling them to blink.

All male lizards and snakes are very well-equipped sexually when compared with mammals. They have two hemipenes which are equipped with all sorts of knobs and bumps - they resemble the exotic 'tickler' condoms that are sold by sex shops. Some lizards eg goannas and skinks, have bones in their hemipenes.

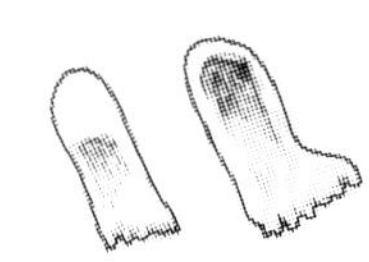

Hemipenal bones

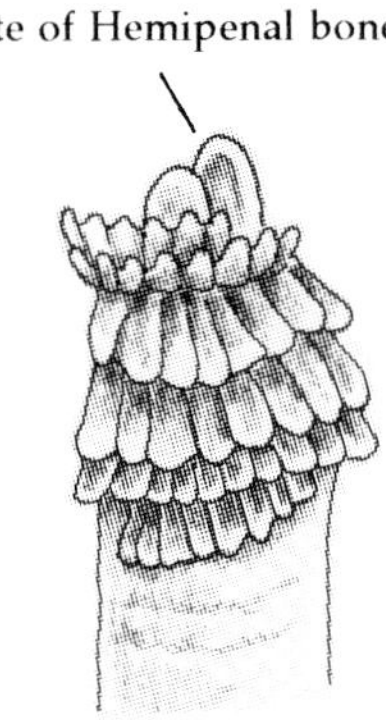

Hemipene of a large goanna

Lizards and snakes can lay eggs , or they can bear live young. In this case the young actually hatch within the mother and are born live. In others, eg Blue-tongued Lizard, the young are attached by a placenta which is eaten after the birth.

There are five families of lizards in Australia and about 480 species.

GECKOES
Family: Gekkonidae

This large family consists of about 800 species which are widespread throughout the world but concentrated in the tropics. There are approximately ninety species in Australia, about fifteen of which are found in the Top End.

Geckoes are nocturnal, soft-bodied lizards, usually less than 10 cm in length, although one species found in Iran reaches 40 cm. They lack eyelids, having instead a clear lens over the eye, which they clean by licking. They can display by waving their tails, suggesting that they have relatively good eyesight.

Geckoes have five digits on each limb. Many have expanded toe pads that have microscopic ridges covered by thousands of little bristles, called setae. These enable geckoes to cling to walls and ceilings and even glass. They roll their toes up and backwards to break the grip. Species can often be distinguished by the pattern of ridges (see picture next page).

Gekkos lay eggs, usually two. Tails are easily lost, but they do regenerate, although without vertebrae.

Any small dull (as opposed to shiny) lizard found active long after dark is likely to be a gecko.

SPINY-TAILED GECKO *(Diplodactylus ciliaris)*

Description: A bizarre-looking gecko covered with short spines. Variable in colour, it may be bright orange and black on a grey background, or pure white with orange and black flecks, or rather dull with white and orange flecks. 8cm (excluding tail). Relatively common. (Picture next page).

Where found: Found in many habitats - caves, limestone, spinifex across much of northern and central Australia. Records from limestone around Mataranka and the tors around Mt Bundey, and the Borroloola region. Recorded also from Cutta Cutta Caves and Manton Dam Reserves and most national parks in the Top End.

NORTHERN DTELLA *(Gehrya australis)*

Description: A plain pale sturdy gecko with large toe pads. Sometimes indistinct bands are present on the back and down the tail. Those who have either good eyesight or new glasses will note that the inner toe is not clawed. 7.5 cm (excluding tail). Relatively common. (Picture next page).

Where found: Found in open forest and savanna woodland, in caves, old mine-workings and also buildings throughout the Top End and into northern Western Australia. Found around Darwin. Recorded from Cutta Cutta Caves, Howard Springs, Manton Dam and Umbrawarra Gorge Reserves and most national parks in the Top End. Other records from Wildman River/Mt Bundey area, Mt Wells, and the Pine Creek and Borroloola regions.

Similar species: Can be confused with the common **House Gecko, *Hemidactylus frenatus*,** brought in from Asia last century. It is sometimes pale and unpatterned, like Northern Dtella, or dark with obvious patterns. It has tubercules or spines along the sides of its tail. This genus has a different pattern of ridges on the underside of its feet, and all toes are clawed. The pads are much smaller than those of *Gehyra australis*. This is the gecko heard barking in Top End homes at night. It can be relatively active during the day. Asian House Gecko is found throught South East Asia, New Guinea and Pacific islands. In Australia it is found in the Top End and also in northern Queensland. It is common in Darwin. There are also records from Litchfield, Kakadu and Nitmiluk National Parks, and Umbrawarra Gorge Reserve, Mary and Wildman Rivers and the Borroloola region.

Toe pads

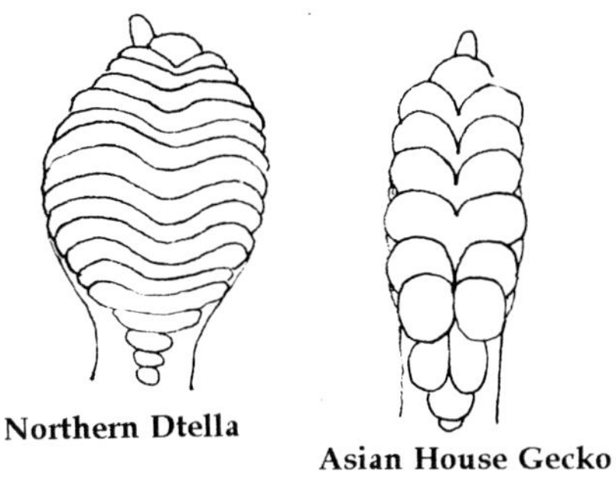

BYNOE'S GECKO/PRICKLY GECKO *(Heteronotia binoei)*

Description: A small variably-coloured gecko (5 cm). Its two most distinctive features are the spines that cover its body, and its bird-like feet. Very common in some areas.

Where found: In a variety of habitats throughout much of Australia. Recorded at Cutta Cutta Caves, Howard Springs and Manton Dam Reserves and most national parks in the Top End. Other records from Mary River, Pine Creek and Borroloola regions.

MARBLED VELVET GECKO *(Oedura marmorata)*

Description: A large pretty cave-dwelling gecko with ragged bands of purplish-brown and cream, paler underneath (this can vary quite a lot. Juveniles are more heavily banded). The tail is usually short and fat in Top End populations. Its feet bear moderately-sized pads. 9 cm, bigger than most Australian geckoes. (Picture next page).

Where found: Found over most of Australia, often in caves and sometimes in open forest. Records from Cutta Cutta Caves, and Umbrawarra Gorge Reserves, most national parks in the Top End, and the Borroloola region.

Notes: *Oedura* is an endemic Australian genus.

Similar species: Giant Cave Gecko *(Pseudothecadactylus lindneri)* is found in Western Arnhemland and northern Western Australia. It can look similar to Marbled Velvet Gecko in colour and shape, but has more crossbands and its tail is slender. It is also slightly larger. Also the ridges of the toe pads extend almost the full length of the toe.

ZIGZAG GECKO *(Oedura rhombifer)*

Description: A tree-dwelling gecko, brown or grey with a zig-zag pattern down the middle of its back.

Where found: Found right across northern Australia. I have seen it on the trunks of felled and live trees in open forest, and under sheets of old iron in the vicinity of Pine Creek. Records from Manton Dam and Umbrawarra Gorge Reserves, and most national parks in the Top End. Also from Douglas and Mary Rivers and at Fountainhead and in the Borroloola region.

DRAGONS
Family: Agamidae

This large family consists of about 230 species found from Europe to Africa, through SE Asia to Australia and into the western Pacific islands. About 50 species are found in Australia.

Dragons, or agamids, are related to lizards found in Africa and Asia. They are known to have been on the Australian continent for at least 25 million years which suggests that they were here when Australia was part of Gondwana. After all, 25 million years ago Australia was in mid-ocean having broken away from that part of Gondwana now known as Antarctica. The earliest agamid fossils have been found in the northern hemisphere and are dated from about 100 million years ago.

Agamids have rough, often keeled scales which are never glossy, and tend to be of similar size all over the body. Their heads are rather blunt and rounded compared with other lizards. Unlike skinks and geckoes they cannot regenerate damaged tails.

Agamids are diurnal lizards. They perform social displays, like arm-waving and males have breeding colours suggesting that they rely heavily on eyesight. They are all egg-layers.

Agamids feed on invertebrates (including spiders), small mammals, and occasionally vegetable matter. They can change colour - dark when it is cool to enable them to absorb heat, and lighter during the day to reflect heat. They also change their posture, raising their body off a hot substrate, to enable thermo-regulation. Any rough-scaled blunt-nosed lizard found active during the day is likely to be an agamid.

FRILLED LIZARD *(Chlamydosaurus kingii)*

Description: A large grey/brown lizard with a large frill usually folded back. The body and frill are often flecked flame-red. 25 cm. (excluding tail). The young look completely different, being much smaller, black and white and grey, and heavily patterned. Common.

Habits: Eats invertebrates but feeds little throughout the dry season when it lives in the treetops. It runs on its hind legs when frightened, usually up the nearest tree, or person, as my friend Penni Tastula, reports with glee. They are sometimes found on the road on cool mornings attempting to catch the sun. When approached it will flatten itself and attempt to look insconspicuous. The frill is erected as a defensive mechanism or in territorial displays.

Where found: Found across northern Australia and down Queensland coast in eucalypt forest and savanna woodland. Hilary Thompson and I found a juvenile in our Parap garden. It used to sun itself in the tops of the trees near the bacony, peering myopically at passers-by. There is one living at Mirambeena Motel in Cavenagh Street, others at Myilly Point, and I often see them on the road to the Sanderson Sewerage Ponds. There are also records from Cutta Cutta Caves, Fogg Dam, Howard Springs, Manton Dam and Umbrawarra Gorge Reserves and all Top End national parks. Can often be seen along the Arnhem, Kakadu and Stuart Highways. during the wet season.

Notes: This is a monotypic genus. Gunwinggu people call it Ngundaman and it is also known as Bemmung (blanket lizard) - this term likens the frill to a man's foreskin. Gunwinggu people, especially the old people, like to eat this lizard. The guts are removed and it is cooked in a hole in the ground.

RING-TAILED DRAGON *(Ctenophorus caudinctus)*

Description: A grey/brown rock-dwelling lizard with a banded tail. It is sometimes spotty or mottled. Male has a crest. 10 cm (excluding tail). Relatively common in suitable habitat.

Habits: an insect-eater. It runs away rapidly when disturbed.

Where found: Found across northern, western and central Australia in rocky areas. It can often be seen hiding under rocks in escarpment country. Records from Arnhemland and Kakadu National Park (found at Little Nourlangie and Nourlangie Rock, and in rocky outcrops in south eastern Kakadu), and also in the Pine Creek and Borroloola regions.

Notes: The best way to see this lizard is to sit quietly among the rocks in the early morning. Often Ring-tailed Dragon will bounce up to sun itself, sometimes approaching within a couple of metres.

TWO-LINED DRAGON *(Diporiphora bilineata)*

Description: Small grey/brown lizard with a long tail and two white stripes running either side of its spine. 7 cm (excluding tail). Common. (Picture next page).

Habits: an insect-eater. Two-lined Dragon is sometimes known as the Ta-ta or Bicycle lizard through its habit of waving its arms in a circular motion. This habit is found in Gilbert's Dragon and Northern Water-dragon.

Where found: Can be found in a variety of habitats including savanna woodland and eucalypt forest Found across the Top End and north Queensland. Top End records from Fogg Dam, Howard Springs, Manton Dam and Umbrawarra Gorge Reserves, and all Top End national parks. Records also from Mary River, Pine Creek and Douglas River and Katherine, Arnhemland and Borroloola regions.

Similar species: See under **Gilbert's Dragon** below.

DIPORIPHORA MAGNA

Description: This lizard comes in all shades of brown. However males turn bright yellow when breeding, with a distinctive black shoulder patch. About 9 cm (excluding tail). Does not seem to be very common.

Where found: Found across northern Australia in grassland and open forest. Top End records at Cutta Cutta Caves Reserve, and Gregory, Kakadu, Keep River, and Nitmiluk National Parks. Also in the Pine Creek and Borroloola regions.

GILBERT'S DRAGON *(Lophognathus gilberti)*

Description: Large tree-dwelling dragon with a white stripe from snout to flank. This stripe covers the lower jaw and a little of the upper jaw but does not reach the eye. 15cm, in length although immatures are much smaller. Adults have a nuchal crest. Juveniles, immatures and sometimes adults may be heavily patterned. Tail and legs bear indistinct bands. Common.

Where found: It is an arboreal species most often found in eucalypt forest and savannah woodland across northern Australia. Tends to be less common than Northern Water-dragon in the northern Top End. Recorded from Fogg Dam, Howard Springs, Manton Dam and Umbrawarra Gorge Reserves, and most national parks in the Top End. In the Pine Creek and Borroloola regions.

Similar species: Northern Water-dragon ***(Lophognathus temporalis)*** looks like Gilbert's and is a similar size. It also has a white stripe from snout to flank, which can continue down the tail as a series of bands. This white stripe almost touches the eye (see picture below). A common species, it breeds in our garden in Parap, sunning itself on the woodpile and in the tops of trees. It often runs into the fruit of the Carpentaria palm *(Carpentaria acuminata)* after insects. Northern Water Dragon is confined to Top End and tip of north Queensland. It is more likely in wetter areas of the Top End. Recorded from Howard Springs, Fogg Dam, Manton Dam and Umbrawarra Gorge Reserves and Gurig, Kakadu, Litchfield and Nitmiluk National Parks. Also present at the Mary River and in the Fountainhead/Pine Creek region.

Two-lined Dragon ***(Diporiphora bilineata)*** is smaller, but can be mistaken for a juvenile Gilbert's or Northern Water-dragon. However this species does not have a white facial stripe.

GOANNAS OR MONITORS
Family: Varanidae

This family is found in Africa, Asia and Australia. In Australia they are known as goannas or monitors. The word 'goanna' is derived from 'iguana' a New World family of lizards that is more closely related to dragons than to goanns.

The oldest fossils of varanids have been found in the northern hemisphere. They date from the late Cretaceous (65-70 million years). Varanid fossils in Australia are much younger, around 23-25 million years. The largest terrestrial lizard ever known, *Megalania prisca,* was related to the monitors. This Australian lizard reached lengths of five metres or more.

The largest member of the family living today, is the Komodo Dragon *Varanus komodoensis*, which grows to about three metres in length. This lizard is found on Komodo and associated islands in eastern Indonesia. They must be a mixed blessing to the villagers on the island, bringing in some revenue from visitors and yet occasionally taking their younger children. Guides armed with sticks escort tourists to the spot where these great lizards are fed. Occasionally one would join the tourists on the viewing platform, causing no little consternation.

The scales of monitors, unlike agamids, are not rough or keeled. They look dull, not glossy like skinks. The scales on top of the head are smaller than those on the rest of the body and the tongue is deeply forked like a snake's. In fact they have a lot in common with snakes, including the tongue, type of teeth and the manner in which they are replaced. Unlike skinks, geckoes and pygopodids, varanids do not drop their tails.

Male monitors are larger than females and are attracted to them by scent. Males have well-developed hemipenes with bones in them (see picture, page 50). Some species of monitor have sex chromosomes - others lack them. All lay eggs.

They have a higher metabolic rate than other lizards of a similar size and tend to rely more on aerobic metabolism rather than anaerobic. This allows them to be active on cool mornings and evenings.

Goannas have been used commercially (eg the famed 'goanna oil' widely used earlier this century for rheumatism). Guinwinggu people eat goanna. They chase it down, grab the tail and hit it with a stick. It is then singed on coals and the entrails are removed, either through the mouth or vent. Cooking time depends on the amount of fat in the goanna. Ten species are found in the Top End.

RIDGE-TAILED/SPINY-TAILED MONITOR *(Varanus acanthurus)*

Description: A small, rock-dwelling goanna with a spiny tail. It is dark or reddish-brown and covered with cream spots or ocelli (eye-like pattern). The head is spotted cream or yellow and there are pale stripes on the neck. There is a dark stripe that runs from the snout, through the eye to the forelimb. Tail is banded dark brown and cream. The short spines are arranged in rings. 25 cm (excluding tail). (Picture next page).

Habits: An insect-eater which may occasionally take small lizards. The spiny tail is used to block the entrance to a hole, thus keeping out predators. I have seen them basking on rocks at Nourlangie and Ubirr Rocks.

Where found: Rocky outcrops through most of northern, central and northwestern Australia.. Top End records from Gregory, Kakadu and Keep River National Parks and Umbrawarra Gorge Reserve. Other records from Arnhemland, Mt Bundey, Pine Creek/Katherine and Borroloola regions

Similar species: ***V. baritji*** also has a spinose tail but is bigger and covered with spots rather than ocelli. It is known from Kakadu National Park and the Pine Creek/Katherine region. Gunwinggu people call this species Biku and say it is sometimes eaten.

V. primordius has a spiny tail and can be of similar appearance. However it is much smaller (only about 12 cm in length, excluding tail), not as heavily patterned, and its tail is not as spiny. There are sparse records from Berry Springs Reserve, near Mount Todd, and Kakadu National Park. Rare.

LONG-TAILED ROCK MONITOR *(Varanus glebopalma)*

Description: A slender, rock-dwelling goanna with a long snout and very long, thin tail It is light to dark grey/brown wth a pale throat and its upperparts are covered with a reticulated pattern. Its limbs are dark and spotted and the tail is dark except for the latter half which is yellow. 30-35 cm (excluding tail). Relatively common but not very visible. (Picture next page).

Habits: this shy reptile feeds on insects. It wedges itself in cracks in the rock when disturbed. Unlike most goannas this species may be active after dark.

Where found: Confined to rocky country throughout Top End and northwest WA. Recorded from Cutta Cutta Caves and Umbrawarra Gorge Reserves, and most national parks in the Top End. Also in Pine Creek and Mt Bundey area.

Notes: Gunwinggu people call this goanna Bonga. It is only eaten by initiated men and older women. It is said that if young children eat it, the goanna will damage their ears.

Similar species: *V. glauerti* is known from Kakadu National Park and the Kimberleys in Western Australia. It looks similar to Long-tailed Rock-monitor but has a strongly-banded black and cream tail. I saw one at the base of the stairs to the Anbangbang Gallery at Nourlangie Rock in May, 1993.

Ridge-tailed Monitor

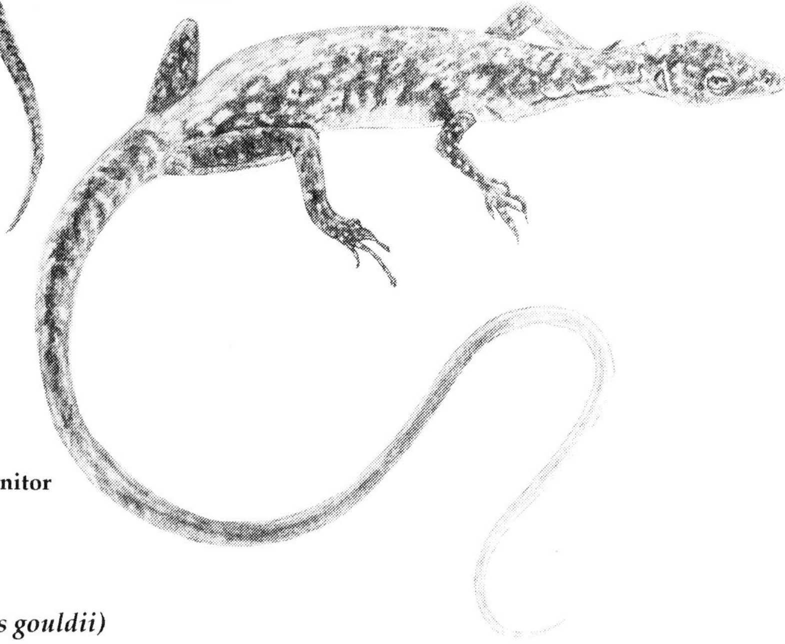

Long-tailed Rock Monitor

GOULD'S GOANNA *(Varanus gouldii)*

Description: A large, common, variable goanna. It can be grey to black, and covered with a variable number of yellow ocelli and spots. Yellow stripes are always present on its face and throat. The legs are also covered with spots and the tail is banded except for the tip which is plain cream. 75 cm in length (excluding tail). Common. (picture next page).

Habits: It feeds on invertebrates, small vertebrates and carrion and often raises itself on its back legs to view the surroundings. It may lay its eggs in termite mounds which makes an idealnursery as temperature and humidity are controlled by the termites. Eggs are laid during the wet season.

Where found: This goanna is found in most of Australia and parts of New Guinea. In the Top End, it has been noted feeding on fledgling pigeons in Fannie Bay and sunning itself by swimming pools. I have seen them dead on Dick Ward Drive and in Parap. Recorded also from Cutta Cutta Caves, Fogg Dam, Howard Springs, Manton Dam, Mary River and Umbrawarra Gorge Reserves, and most national parks in the Top End. Also the Pine Creek and Borroloola regions.

Notes: They are often handfed by visitors in places like Fogg Dam. Some can become quite aggressive towards people, possibly as a result of this. DO NOT FEED.

They seem to have marvellous powers of recuperation. A large Gould's goanna was badly injured in the back of the skull at Howard Springs. Rangers asked me to take it although we all agreed it would probably die. The goanna was completely paralysed but after a few days of feeding with a special recipe (including that wonder food Vegemite), it suddenly regained use of its legs completely, performed a death-defying leap through the window, and scurried off down the street towards the bush. Gunwinggu people call this species Djani and say 'it is good for eating'.

Similar species: ***V. panoptes*** looks similar but has alternating rows of dark and light spots that continue onto the tail, the tip of which is banded. I found one dead on the road to the Northern Territory Museum. about May, 1993. Paul Horner says that this species is more common in Darwin than *V. gouldii*. It is found across northern Australia and in parts of Western Australia. Recorded from Fogg Dam, Manton Dam, Mary River and Umbrawarra Gorge Reserves, and Gurig, Kakadu, Keep River and Litchfield National Parks. Also found at Wildman River and in the Pine Creek and Borroloola regions. My adopted Gunwinggu son says that this is a 'ceremony animal' and is only to be eaten by men and women who 'pass ceremony tests'. He calls it Garawan.

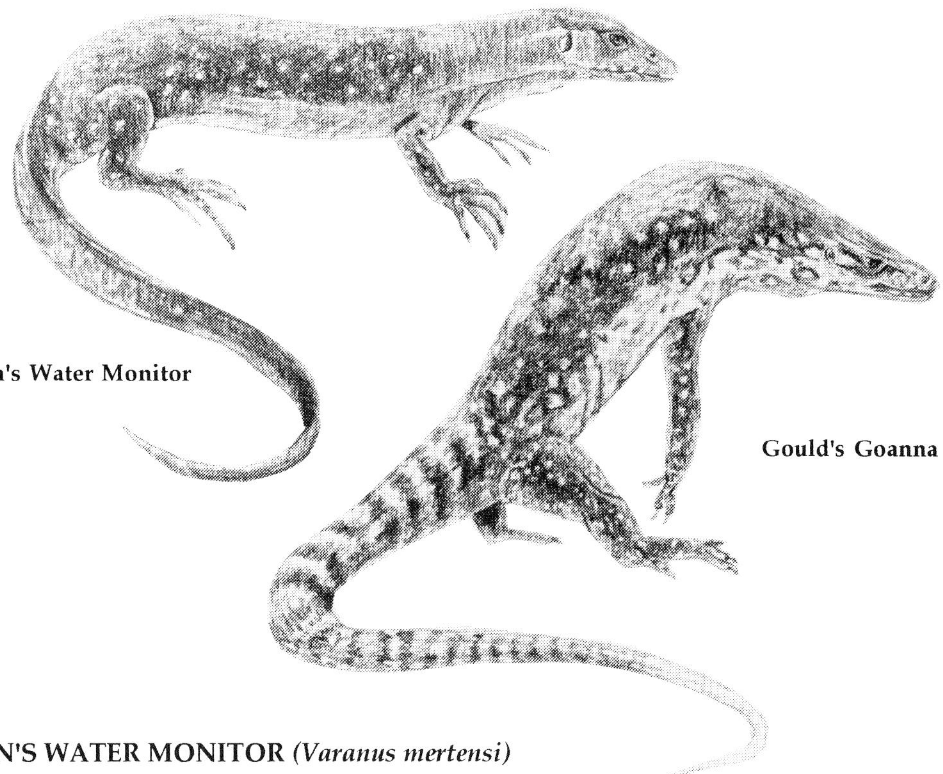

Merten's Water Monitor

Gould's Goanna

MERTEN'S WATER MONITOR *(Varanus mertensi)*

Description: large plain dark-grey varanid covered with tiny yellow spots. Tail is laterally compressed. 0.75 m (excluding tail) but is often much smaller. Common.

Habits: This species normally eats fish but has learnt in some areas to take a variety of food from visitors. This and the other aquatic species mentioned in this book, *V. mitchelli* and *V. indicus*, breed in the early dry season.

Where found: Near fresh water across northern Australia. Recorded from Darwin, at Casuarina Coastal Reserve. Also Berry Springs, Butterfly Gorge, Cutta Cutta Caves, Fogg Dam, Howard Springs, Manton Dam, Mary River Crossing, Umbrawarra Gorge Reserves, and most national parks in the Top End. Other records from Douglas, Daly Rivers, Mary and Wildman Rivers. Also found in the Pine Creek and Borroloola regions. One joined us in the hot springs off the Daly River Road, while we swimming.

Notes: These can become very aggressive when semi-tame and will attack. DO NOT FEED. Like other large monitors they can inflict nasty bites and scratches. Wounds can quickly become septic if not treated. Gunwinggu people call this species Burrbu and say it is 'good to eat'. They cook it in a hole in the ground.

Similar species: ***Merten's Wate Monitor*** is plain compared with Gould's Goanna and Mitchells Water Monitor, and is much larger than the latter.

Mangrove Monitor *(V. indicus)* is more heavily spotted and smaller. It is found in monsoon forest and mangroves in the Top End, in northern Queensland, and in New Guinea.

MITCHELL'S WATER MONITOR *(Varanus mitchelli)*

Description: A small dark grey or brown goanna with numerous spots and ocelli. The head and throat are spotted and striped with yellow. The dark tail has a few yellow flecks but is basically plain especially toward the tip. Tail laterally compressed. 30 cm (excluding tail).

Where found: Basically in the same aquatic habitats as Merten's Water Monitor throughout the Top End and northern Western Australia. Darwin record from Sanderson Sewerage Ponds. Records also from Manton Dam, and Umbrawarra Gorge Reserves, and Kakadu, Keep River, Litchfield and Nitmiluk National Parks. Also found on Marrakai Plains/Mary River wetlands and in the Pine Creek and Borroloola regions.

Mitchell's Water Monitor

Spotted Tree Monitor

SPOTTED TREE MONITOR *(Varanus scalaris)*

Description: A variable, little goanna, pale rufous, grey or black with white or cream ocelli and spots. Throat is white or cream and underparts are white. Tail is often banded but is sometimes plain black, and has strongly keeled scales. 25 cm (excluding tail).

Habits: Basically this species is an arboreal lizard but I have sometimes caught them in Elliott traps placed on the ground, and it is sometimes killed on the roads. It feeds on insects and small vertebrates.

Where found: Found all over the Top End and across northern Australia, in eucalypt woodland and open forest. Top End records from Darwin, and Fogg Dam, Howard Springs, and Manton Dam Reserves, and most Top End national parks. Other records from Mary River/Mt Bundey and Douglas River/Pine Creek and Borroloola regions.

Similar species: Freckled Monitor *(Varanus tristis)* looks similar with strongly keeled scales on its tail. However the head, neck and tail may be black, although often the head has a broken reticulated pattern. The head of *V. scalaris* has a relatively blunt snout whereas *V. tristis* is more pointed. The supraocular scales (those above the eyes) are very small compared with those of *V. scalaris*. Also *V. tristis* males have visible spines on either side of the vent (see picture below). It grows a little bigger than *V. scalaris* and its tail is longer *V. tristis* is arboreal and can also be found in rocky outcrops. Found over much of Australia. Top End records from Arnhem Highway, Blackmore River and Victoria Highway. Recorded also from Cutta Cutta Caves Reserve and Kakadu, Keep River, Litchfield and Nitmiluk National Parks. Also found around Pine Creek and Borroloola regions.

Head scales of Spotted Tree Monitor

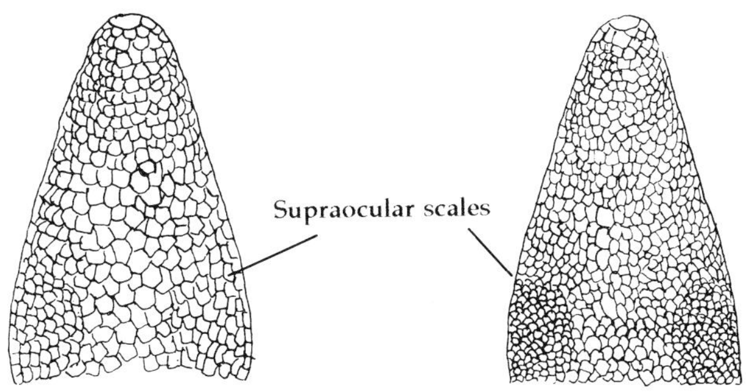

Head scales of Freckled Monitor

Vent of Freckled Monitor

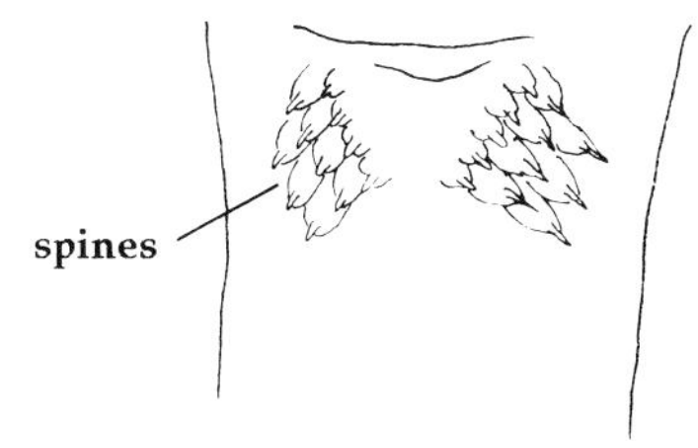

LEGLESS LIZARDS
Family Pygopodidae

This family is confined to Australia and New Guinea. Most are not very noticeable and consequently only one species has been included.

They lack any sign of forelimbs, the hindlimbs only represented by a flap near the vent. Some species have an external ear opening. They lack eyelids, like snakes. Like snakes and elongated skinks some of their internal organs have become elongated or reduced, eg the lungs. Females are usually larger than males and are egg-layers. Like skinks and geckoes they lose their tails easily.

BURTON'S SNAKE-LIZARD *(Lialis burtonis)*

Description: A legless lizard with a pointed snout. It is variable in colour and may be plain, striped or spotted. At 30 cm (excluding tail) it is the largest member of this family. It has the ability to vocalise, like geckoes, and its snout can bend at a joint between the eyes, a device that may help it hold its slippery prey - small skinks, more easily. Its teeth are slightly recurved, more like those of snakes than other pygopodids which usually have peg-like teeth. Relatively common.

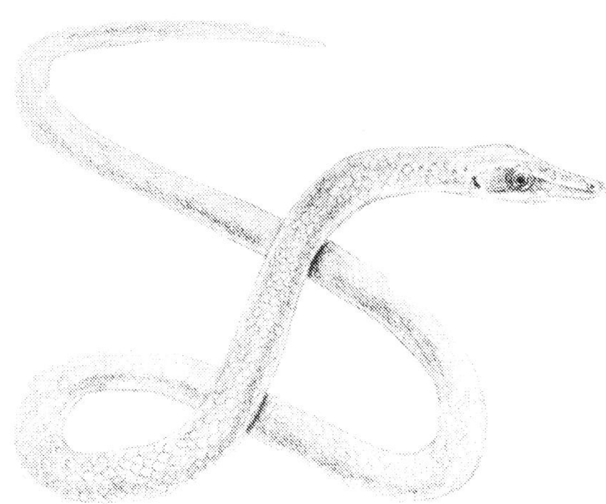

Burton's Snake-lizard

Habits: Feeds mainly on skinks which it ambushes by waving the tip of its tail in the air to attract them, and then pouncing.

Where found: Found throughout Australia and in New Guinea in a range of habitats from monsoon forest to savanna woodland. Records from Cutta Cutta Caves, Howard Springs and Manton Dam. Reserves and most national parks in the Top End. Also Daly River and the Pine Creek and Borroloola regions.

SKINKS
Family Scincidae

This is the largest family of Australian lizards. Most have smooth shiny scales, those on the head being larger than those on the body. Some have moveable eyelids, in others the eyelids are fused into a transparent scale, much as in snakes.

Skinks are mainly diurnal and can often be seen running around in the heat of the day. They prevent overheating by moving from light to shade. Few skinks develop breeding colours, suggesting that their eyesight is poor. Some lay eggs and some bear live young, a 'true' placenta developing in some of the latter. Females are often larger than males.

Only those that the local or visitor is likely to notice are included here. For a complete rundown on Northern Territory skinks look to Paul Horner's book *'Skinks of the Northern Territory'*.

FOUR-FINGERED/JEWEL/RAINBOW SKINKS *(Carlia* species)

There are 24 Australian species in this genus and six in the Top End. This genus extends into New Guinea and eastern Indonesia.

Unlike most other skinks, males develop beautiful breeding colours hence one of their common names. Non-breeding males, and females are difficult to identify. These lizards display, waving their tails and bobbing their heads. They are both terrestrial and diurnal, although, in the Top End, some are active after sunset in the hotter parts of the year. Breeding occurs during the wet season. All species are egg-layers.

Top End species are about 3.5 - 4.5 cm long (excluding tail).Three of the most common species are detailed below. All are basically brown above with white underparts.

CARLIA AMAX

Description: A tiny, black-flecked skink commonly found in or near rocky areas. It has brown streaks on its throat and a short, fine white line just under the eye. Dorsal scales (for those keen observers with a hand lens) are six-sided with two keels and an angular free edge. Breeding males have a copper-green head and red flanks. Common in suitable habitat.

Where found: in rocky areas and hills throughout the Top End and northern Western Australia. Top End records from Cutta Cutta Caves, Howard Springs, Manton Dam and Umbrawarra Gorge Reserves and most Top End national parks. Other records from Douglas and Mary River and the Mt Wells/Pine Creek and Borroloola regions.

Similar species: ***Carlia gracilis*** is plain brown above (no black flecks) without an obvious white line under the eye and it is found in wetter areas than *C. amax*, eg monsoon and gallery forest. Dorsal scales are similar to those of *C. amax* but have three keels. Breeding males develop copper colours on the body and forelegs with a red stripe along the side and irridescent blue-green on the head. A common species in suitable habitat. Recorded from Fogg Dam and Umbrawarra Gorge Reserves, and Kakadu, Keep River and Litchfield National Parks. Other records from Douglas, Mary and Wildman Rivers and the Pine Creek and Borroloola regions. Found throughout the coastal Top End, including Darwin gardens. Common.

Carlia munda is more heavily flecked than either *C. amax* or *C. rufilatus* ,and has a relatively prominent broken white line extending from snout to past the forelimb. Dorsal scales may or may not have keels (usually three weak keels if present) and have a curved free edge. Breeding males develop a red stripe along the side. This species is found across northern Australia including the Top End, with records from Darwin and Bathurst Island, and

Gurig, Kakadu, Litchfield and Nitmiluk National Parks and the Pine Creek and Borroloola regions. Common.

Carlia rufilatus is flecked with black like *C. Amax* and also has a fine white stripe from below the eye but in *C. rufilatus* this extends to the forelimb. Breeding males have a red mid-lateral stripe. Dorsal scales are just like those of *C. gracilis*. Found in the north and west of the Top End and in northern Western Australia. Records from Darwin and Gurig, Kakadu and Litchfield National Parks. Common.

SNAKE-EYED/WALL SKINK *(Cryptoblepharus plagiocephalus)*

Description: A slender, patterned, tree-dwelling skink with five digits on each foot, found in local gardens and elsewhere. 45 mm (excluding tail). It is usually greyish with a ragged pattern down the middle of the back. These lizards are called snake-eyed because like snakes, they lack movable eyelids, having instead a transparent scale over the eye. Common. (Picture next page).

Where found: This arboreal skink is found over most of Australia. Recorded from Darwin and Bathurst Island, Cutta Cutta Caves, Fogg Dam, Howard Springs, Manton Dam and Umbrawarra Gorge Reserves, and most Top End national parks. Other records from Fountainhead/Pine Creek, Katherine and Borroloola regions and Douglas, Mary and Wildman Rivers.

Similar species: *Cryptoblepharus carnabyi* looks similar. To tell them apart look at the plantar scales (on the sole of the foot) with a hand lens. *C. plagiocephalus* has rounded plantar scales, *C. carnabyi* has pointed scales. A wall skink in a Darwin garden is likely to be *C. plagiocephalus*. *C. carnabyi* is found over much of coastal Australia. I have a Top End record from the Douglas River area and there is a Conservation Commission record from Kakadu National Park. Paul Horner records this species from the Gulf of Carpentaria region.

Another species, *C. megastictus* is sometimes seen on the rocks at Nourlangie and Ubirr Rocks in Kakadu National Park. It is a rich brown above with white and dark brown spots and streaks. Also found in the Borroloola region and into the Kimberleys in Western Australia.

Notes: I have seen Magpielarks, *Grallina cyanoleuca,* catch and eat Wall Skinks. The genus *Cryptoblepharus* is also found in Africa, Indonesia and New Guinea. They are all egg-layers.

CTENOTUS SPECIES.

There are about ten species of this genus in the Top End.

Description: *Ctenotus* species are bigger and bulkier than any of the lizards described above, and they are usually have prominent stripes and/or spots. Any medium-sized striped and/or spotted skink with well-formed legs is likely to be a *Ctenotus* species, if it does not have a plain red tail. They also have prominent lobules on the ear opening and are sometimes called Comb-eared Skinks because of this feature.

They are found in most habitats, usually on the ground or rocks. However they are not a prominent feature of Darwin gardens. Members of this genus can sometimes be seen on the rocks at the art sites at Nourlangie Rock and Ubirr in Kakadu National Park. They are present in most parks and reserves. One species is described below.

PORT ESSINGTON SKINK *(Ctenotus essingtoni)*

Description: Medium-sized skink with a plain brown/grey back and a broad black stripe on each side bordered by white stripes. The lower sides are dull rufous with white spots. 6 cm excluding tail. Relatively common. (Picture below).

Where found: In coastal and sub-coastal open forest and woodland across the northern Top End and northern Queensland. Recorded from Darwin, Fogg Dam and Howard Springs Reserves, and Gurig, Kakadu, Litchfield and Nitmiluk National Parks. Also found around Fog Bay and Pine Creek.

Notes: named after Port Essington, an early European settlement made last century on Cobourg Peninsula.

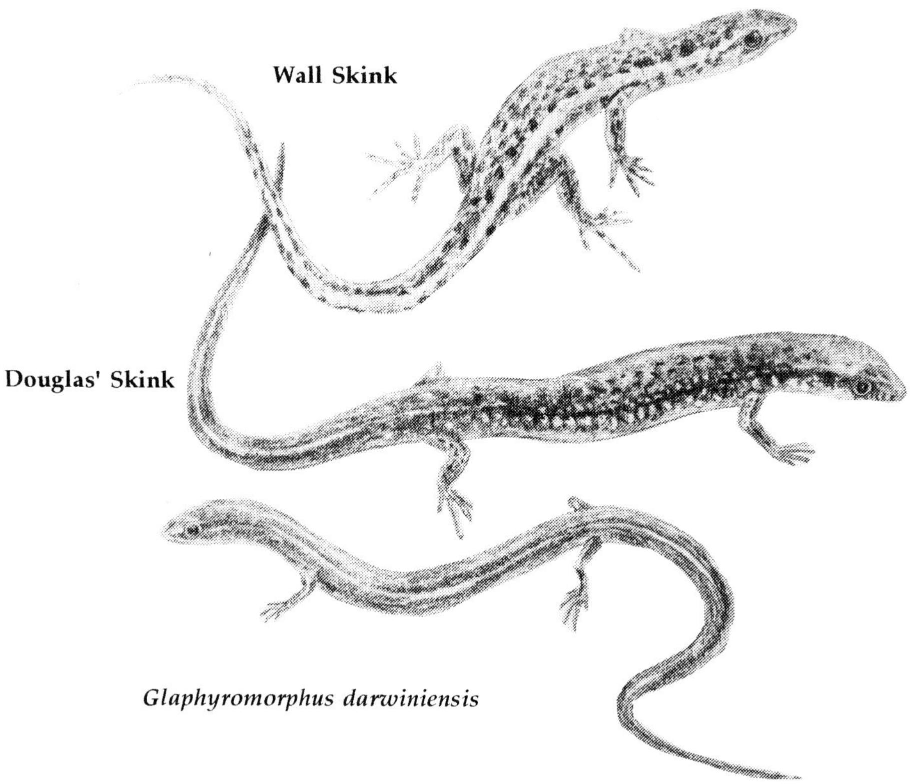

DOUGLAS' SKINK *(Glaphyromorphus douglasi)*

Description: An attractive sturdy, small-legged skink found in local gardens. It is basically brown with darker flecks above and white spots (not always present) along its flanks which have a pink or rufous tinge. 8 cm (excluding tail). Common.

Habits: Usually eats invertebrates in the wild but has been noted scavenging food scraps and nibbling the dog food in my Parap garden. It topples into my fishpond occasionally, providing me with lots of study specimens. However it always saddens me to see their plump little bodies bottom-up.

Where found: Frequents leaf litter near riverine habitats throughout western Top End, often found in Darwin gardens eg Parap where they breed. Recorded from Fogg Dam and Howard Springs Reserves and Gurig, Kakadu, Litchfield and Nitmiluk National Parks. Other records from Mary River, Borroloola and Pine Creek regions. This is a large genus found throughout SE Asia and into New Guinea and the Pacific Islands.

Similar species: Can be confused with *Ctenotus* species but is not as heavily marked, and does not have ear lobules. ***Glaphyromorphus darwiniensis*** is a very elongated small-legged skink also found in Darwin gardens. They are dark brown and rather plain compared with *G. douglasi* and much more cryptic -they hide in leaf-litter, and as I've found, dirty linen left on the laundry floor downstairs! Fortunately I've not washed any yet!

FIRE-TAILED SKINK *(Morethia ruficauda)*

Description: A distinctive black skink with broad white stripes and a red tail. 5-6 cm. (excluding tail). Common.

Where found: All over the Top End, including Darwin gardens. where there are records from Parap and the northern suburbs. Can be seen running around on the rocks at Ubirr and Nourlangie Rocks in Kakadu National Park. Recorded also from Butterfly Gorge, Cutta Cutta Caves, Manton Dam and Umbrawarra Gorge Reserves, and most national parks in the Top End and the Pine Creek area.

Notes: *Morethia* species are distinctive in the Top End (although not necessarily elsewhere in Australia) because they are all black/grey and white-striped with bright orange/red hindlegs and tails. Like *Cryptoblepharus* species they have an immovable lens over the eye and five digits on all feet. They are egg-layers.

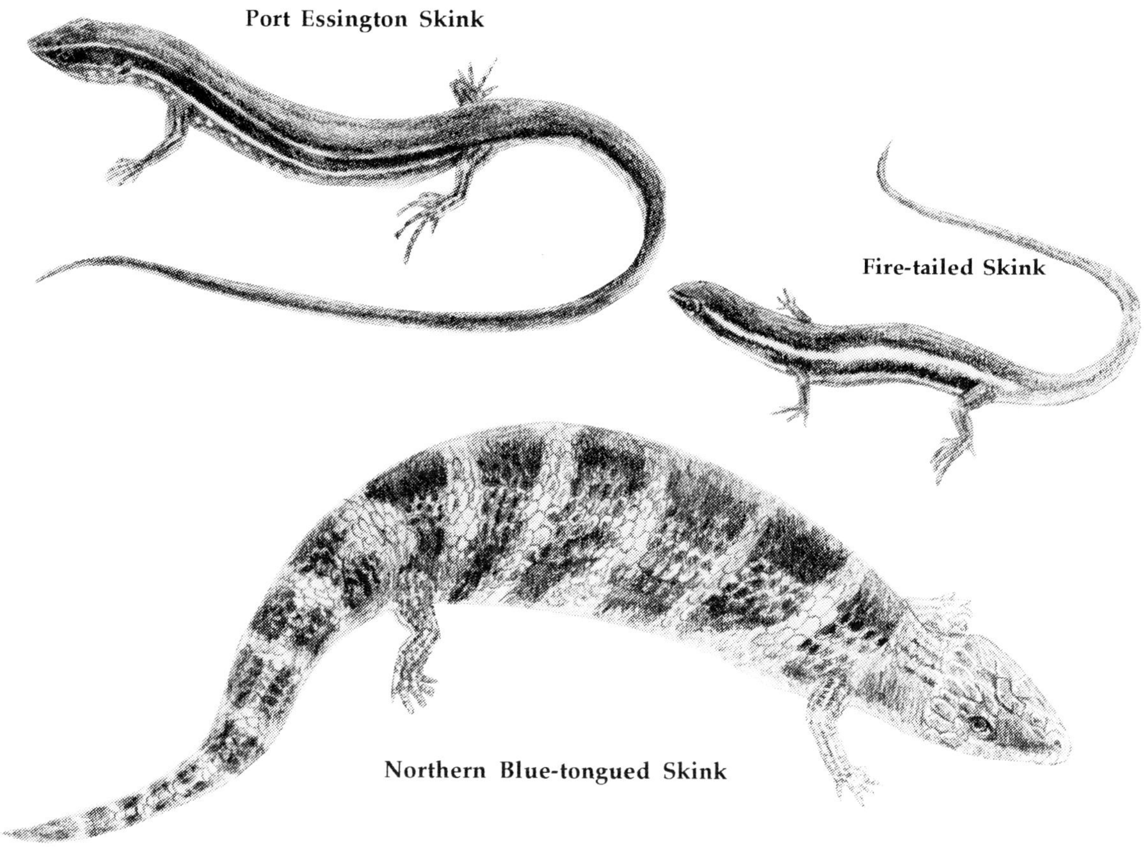

NORTHERN/COMMON BLUE-TONGUED LIZARD *(Tiliqua scincoides)*

Description: A very large, thickset skink with a broad head, dark blue tongue and small limbs. It is usually grey with broken bands of yellow, brown and/or black, which may not extend to the tail. The tail is thick and not fragile (meaning it does not fall off when this lizard is attacked). 30 cm (excluding tail). Southern populations have a dark streak behind the eye. Common.

Habits: When this lizard is threatened it exposes its tongue and hisses loudly. It naturally feeds on invertebrates and vegetable matter. However in our garden it steals eggs and fowl food to the disgust of my pet chickens, and dog food. They bear live young.

Where found: This lizard is found throughout northern and eastern to south-eastern Australia. It is found all around Darwin where and often falls prey to cats. It breeds in my garden in Parap.

Recorded from Cutta Cutta Caves, Fogg Dam and Manton Dam Reserves, and most national parks in the Top End. Other records from Douglas, Daly and Mary Rivers, at Fountainhead and Katherine.

Notes: Gunwinggu people call this lizard Gurrdirr and it is eaten.

Similar species: Centralian Blue-tongued Lizard *(T. multifasciata)* is also found in the Top End eg records from Kakadu and Nitmiluk National Park, and along the Victoria Highway into Gregory National Park. It has a dark mark reaching behind the eye, as in southern populations of Northern Blue-tongued and 2-3 supra-oculars (scales above the eye), as opposed to four in the Northern Blue-tongued. Anterior temporal scales (scales behind the eye) are smaller than in Northern Blue-tongued and its tail is shorter.

SNAKES

Order Squamata
Sub-order Serpentes

There are five families of terrestrial snakes found in Australia, Tylophidae, Colubridae, Acrochordidae, Boidae and Elapidae. The Elapidae, to which cobras, coral snakes, mambas and kraits belong, is the dominant family of snakes found in Australia. The family Colubridae dominates everywhere else. In fact three quarters of all species of snakes are colubrids.

Snake fossils are known from the Cretaceous period (130-65 million years ago). The oldest Australian snakes probably originated in Gondwana, the supercontinent that separated into the countries of South America, Africa, Iran, India and Australia, as fossils of matsoiid snakes, giant python-like creatures, found here, are related to fossils in Patagonia, Madagascar and Egypt. Elapids may have colonised Australia from SE Asia 13 million years ago. The fossil record for these snakes, and also for colubrids, and achrochordids (file snakes) is quite poor in Australia.

Snakes have similar internal organs to other animals. However one of the paired organs may be missing, some such as the lung are elongated, others like the kidneys (which are not kidney-shaped) and testes are staggered - one is in front of the other. To save room the intestine is only slightly coiled. Snakes have no bladder.

The anatomy of a snake

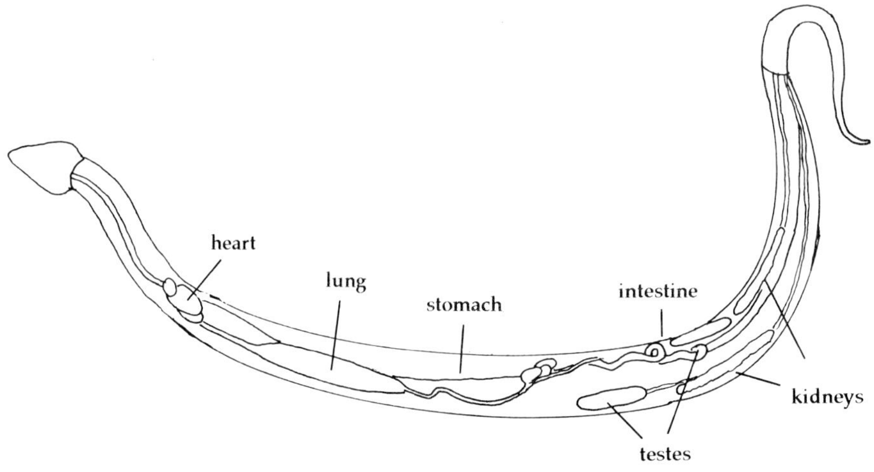

A point of differentiation between lizards and snakes is that the latter have ligaments between the two halves of the lower jaw, which allow them to move independently of each other. This, and their lack of a sternum, enable them to eat big prey. Snakes are also able to protrude the epiglottis which enables them to breathe while their mouths are full.

Like lizards, male snakes have two hemipenes which are often well equipped with spines and other paraphenalia. Males appear to find receptive females by scent which the female produces from anal glands. Some snakes actually indulge in foreplay. For instance a male may rub a female's head or chin. Python males may caress a female with his cloacal spurs (see Pythons, page 70). Females often move off during mating, dragging the males behind, possibly why they have the spines. Snakes may lay eggs or bear live young. Rival males often indulge in wrestling.

SAFETY

Snakes are often regarded with great fear. Yet many are harmless and few are interested in attacking humans. Many victims are bitten when they step on snakes, often in the cool of morning or night when snakes are rather sluggish. Others are bitten when they put their hands into hollow logs or leaf litter, or when they attempt to kill snakes.

Most venomous Australian snakes have minute fangs. Because of this, and the fact that most bites do not cause local tissue damage, people may not even know that they have been bitten until they start showing symptoms. A woman friend of mine was bitten while in her yard. She saw the snake but didn't see it bite her. About ten minutes later she felt distinctly unwell. Later, in hospital, doctors discovered that she had been bitten by an elapid.

Many people cannot tell the difference between venomous and non-venomous snakes.* Therefore it is best to leave all snakes alone. However if you do have a close encounter with one and wish to know in a big hurry whether it was dangerous, note that any snake with big scales (8 -10 across the width of the back) is possibly an elapid and therefore venomous.**

Any snake with small scales (15 -30 across the back) is likely to be a python and therefore relatively harmless. You do not need to count them. They will just look very small, like those depicted below. Also most elapids are rather snub-nosed (see page 80) compared with pythons which are rather dog-faced (see page 71). Incidentally small snakes can be hard to identify.

Snake scales

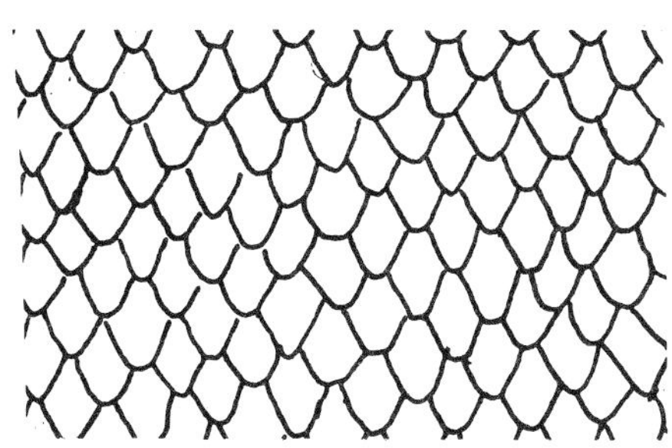

Elapid

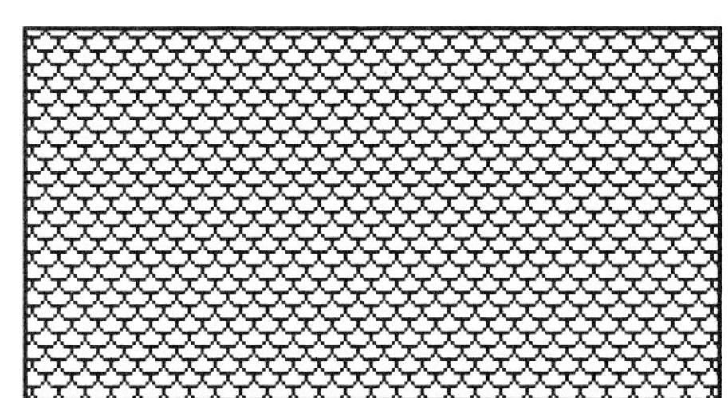

Python

*There is a tale of a King Brown being handed around a hotel bar in the Top End. Those present all thought that it was a python.

**Usually biologists do a full mid-body scale row count - all the scales around the body, as you will note in the descriptions of snakes that follow. The scale count across the back is only done here to enable the layperson to distinguish between possible venomous elapids and non-venomous pythons in a hurry.

Precautions: A few precautions will usually protect you. Wear boots, gaiters or long trousers when walking (I wear gumboots which are cheap and great protection from grass-seeds as well). Carry a broad elastic bandage with you. It is no use to you sitting in the car glovebox. Do not go crawling into caves. I once worked with a biologist who came nose to nose with a King Brown in a mine. He was lying prone upon his belly. Luckily the snake was rather laid back about it all. When I crawl into a cave or mine (as part of my survey work), I check well in front with a torch.

First Aid : Most Australian elapids have rather inefficient fangs and they may not be able to envenomate the victim unless they can deliver two or three bites, or hang on and chew. So chances are, if you or a companion do get a passing bite from a snake, that you have escaped with nothing more than some fine puncture marks and a nasty scare.

However if the worst happens, first signs of envenomation may come within ten minutes -symptoms include sagging eyelids, salivating, difficulty in swallowing, swollen, and painful lymph nodes in the armpit or groin.

There's no need to panic. That makes things worse for everyone concerned. Just wrap a broad bandage firmly around the bite site and then up the limb (almost all bites occur on the lower leg). Venom is usually spread via the lymphatic system and this firm pressure is enough to control its spread. The victim is not likely to die once the bandage is in place properly, even if the hospital is some hours away.

Do not wash the bite as doctors can identify the species of snake from the venom. Washing the bite may be costly as doctors may then have to use a poly-antivenene, much more expensive than a specific one.

Put victims in a position where they can breathe easily, and keep an eye on them. Comfort them by holding a hand, stroking their hair, and singing. The more relaxed the patient, the less rapidly venom will spread. Most people have some idea of how to calm a scared child. Use that knowhow now. Apart from the broad bandage and medical attention, it is the best thing that you can do.

It may make you feel better to know that most Australian snakes do not cause the horrific tissue damage and gangrene associated with viper bites. A substantial portion of people bitten by vipers in the United States, for instance, lose large slabs of tissue, even limbs. In Florida, 17% of childhood amputations are caused by viper bite. There are no vipers in Australia.

First aid for snakes: If you find an injured snake on the road, take care. I have seen badly injured elapids (pythons as well) launch themselves at onlookers. Even if the snake seems near death, be careful. Venom leaks out of fangs when snakes are dying or dead and as they often twitch spasmodically, a careless examiner could end up taking an unexpected trip to the hospital.

I once revived a drowned python by placing it head down on the warm bonnet of a car. It worked. I've never had to try mouth to mouth resuscitation but I once heard of a breeder of rare giant moths who found one of his precious caterpillars drowned in its water bowl. He immediately commenced resuscitation through the spiracles, the little holes in the sides of insects through which they respire, and revived his caterpillar. Therefore there is hope for snakes. Just be advised not to try it on anything venomous.

You can stabilise rib fractures on a python by cutting off the bottom and neck of a plastic bottle. Pad it, slide it up the snake's body and tape it in place (this works for human limb fractures as well). Then slide a board, under the snake and carry it to your vehicle. Berry Springs Wildlife Park takes such injured animals. If you cannot help a badly injured python then dispatch it. The best way to do this is by cutting its head off, or by dropping a large rock on its head. If you think this sounds dreadful then imagine the snake dying slowly while the kites pick its insides out. However you may have no choice but to leave an injured elapid. I have killed them by throwing a rock from a safe distance and thus breaking the spine near the neck. Once immobilised I then dispatch them with a rock or a stick. However I wouldn't advise people to do this as a matter of principle.

Prevention: Bush lore says that snakes do not like geraniums and urges that people plant them around the house. This may or may not work down south but to my knowledge geraniums do not do well in the tropics and people would be better advised to take other precautions especially on a bushwalk when carrying pots of flowers is impractical.

Respect our snakes: Some locals, tour-guides among them, think it's a really macho thing to catch snakes to show people. Unfortunately it may create the impression that venomous snakes aren't really dangerous, or that snakes are just playthings.

Imagine how you'd feel if a tiger approached you. You'd expect it to eat you, not play with you. Wild animals have similar expectations. They are often terrified when caught. They do not know that the nice tour guide is only going to show them to his clients. In this day and age many visitors are put off by such behaviour, believing that wildlife should be left alone. It serves such guides right when a scared snake defecates copiously down their neatly-pressed shirts.

Guides can show visitors a snake on a road by working out its critical distance and walking first to its side, and then into its path, to slow its progress. This way of viewing snakes is relatively non-intrusive and is more acceptable to nature-loving tourists.

Gunwinggu people call all snakes Ngayin.

BLIND SNAKES
Family: Typhlopidae

Blind snakes often turn up in Darwin gardens, Kakadu National Park and elsewhere throughout the Top End. These snakes look much like a shiny worm, although they are not slimy. Tiny black eyes are discernable and there is a spine at the end of the tail which can give your finger a tiny prick. They are completely harmless, except to ants and termites which they eat.

There are over thirty species in Australia, ten of which are found in the Top End. Many are very difficult to identify without a microscope.

Blind snakes, unlike most others, lack enlarged ventral scales. Only the upper jaw has teeth, and there is a single lung. All Australian species are egg-layers.

PYTHONS
Family Boidae

Pythons are found in Africa, Asia, Australia, New Guinea, and in the western Pacific Islands. They belong to the sub-family Pythoninae. Members of the other sub-family Boinae, are found in all of these areas, except Australia.

Pythons can be distinguished from venomous elapids by their elongatedheads, and the small scales - they may have over 65 rows of scales at the mid-body. They also have one or more loreal scales, between the eye scales and the nasal scale. (See illustration of Water Python, page 71).

All, apart from the genus *Aspidites,* have heat-sensitive pits in the scales of the lower jaw (vipers also have heat-sensitive pits) and all have a single anal scale. They often have a mixture of single and divided subcaudal scales (scales on the underside, below the vent). Pythons are egglayers and females protect eggs by coiling around them. a degree of maternal care not shown by other snakes.

Males and females have cloacal spurs. The male's are bigger than the female's and are used to titillate her in preparation for mating. This cloacal spur consists of remnants of hindlimbs - a femur and claw.

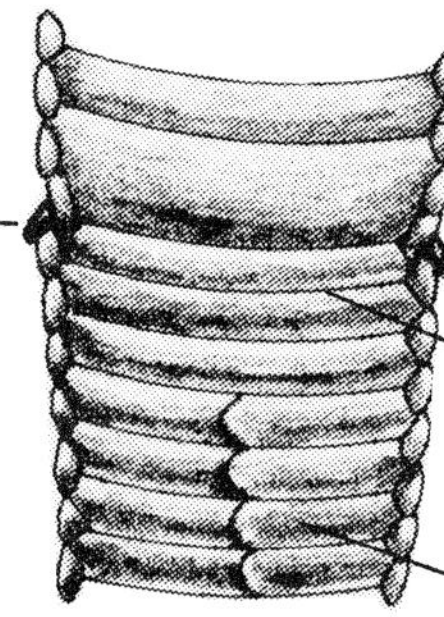

There are about thirteen species of python in Australia. They are concentrated in the northof the continent, eg six species are found in the Top End and none are found in Tasmania.

BLACK-HEADED PYTHON ***(Aspidites melanocephalus)***

Description: A nocturnal, brown snake with darker mottling or bands across most of its body, and a black head and neck. 2 m. Scales 50-65 rows at mid-body. It lacks heat-sensitive pits on the lower labial scales (scales on lips of lower jaw). Relatively common.

Habits: This python eats mammals, birds and reptiles. It lays about ten eggs, a small clutch compared with other pythons.

Where found: A variety of habitats, including open and closed forest, across northern Australia. Top End records from Cutta Cutta Caves, Fogg Dam and Umbrawarra Gorge Reserves and Kakadu, Keep River and Nitmiluk National Parks. Also noted at the township of Adelaide River, at Fountainhead, Pine Creek, the Stuart Highway near Katherine and the Victoria Highway near Gregory National Park. Also records from the Marrakai Plains and the Borroloola region.

Note: This genus is only found in Australia. The only other species is *Aspidites ramsayi,* from arid Australia. The first snake my son Rowan ever held was a large Black-headed Python that I was releasing. I hesitated, as it was a particularly bad-tempered snake and he was only eighteen months old. However Rowan insisted, holding out his arms for me to drape the snake over. It lay quietly coiled around him for a minute, then slid slowly down his back and into the bush.

Children's Python

Black-headed Python

CHILDREN'S PYTHON *(Liasis childreni)*

Description: A pretty, inoffensive brown snake covered with dark purplish-brown blotches which have an irridescent sheen in torchlight. It moves more quickly than most other pythons. 1 m. It has two or more loreal scales. Mid-body scale row count is 37-49. Like most Australian pythons it has heat-sensitive pits. Common. (Picture previous page).

Where found: Found over much of northern Australia (except for eastern Queensland. Top End records from Darwin at Sanderson Sewerage ponds (in June). Records from Cutta Cutta Caves, Fogg Dam and Manton Dam Reserves and most Top End national parks. Other records from Mary River , Pine Creek and Borroloola regions.

Notes: This snake is named after a Mr. Children, who worked for a British Museum, and not because it is of a suitable size and nature for a child's pet.

Similar species: Carpet Python *(Morelia spilota)* is a variable snake but is often similar in pattern and colour to Children's Python. It has numerous loreal scales, 40-65 scale rows at mid-body. Found over much of Australia except the arid regions. Recorded from Darwin, Fogg Dam, Gurig, Kakadu and Litchfield National Parks.

WATER PYTHON *(Liasis fuscus)*

Description: Beautiful, irridescent olive-brown snake, yellow or apricot underneath, with only one loreal scale. It is found near water. Average length about 2 m. with a maximum of about 3 m. Mid- body scale count 40-55 rows. Common in suitable habitat.

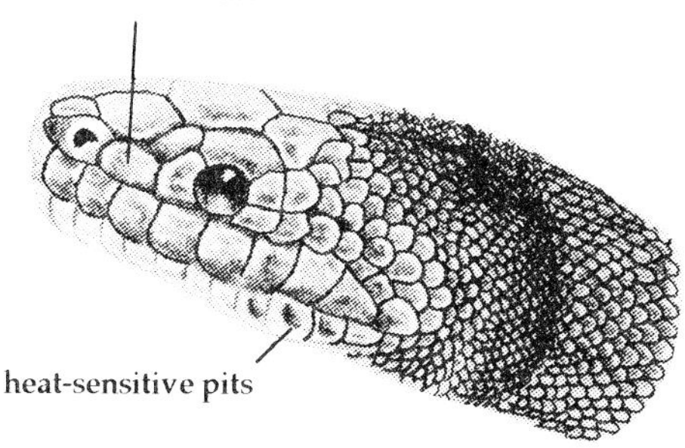

Water Python

Habits: Largely nocturnal although often active early morning and late afternoon. It feeds on frogs and rats. Brreeding occurs around August-September.

Where found: This snake is found in wetlands and creek systems throughout northern Australia and in New Guinea. Top End records from Darwin, and Fogg Dam, Manton Dam and Umbrawarra Gorge Reserves, and most Top End national parks. Other records from Beatrice Hill, Mary River, through Arnhemland and on Bathurst and Melville Islands.

Notes: Some Aboriginal people believe this is the Rainbow Serpent. Gunwinggu people call Bolochor (meaning 'quiet snake'). Do not be fooled by this name. Water Python will happily have a chomp at you if annoyed. Tiwi women do not eat the snake unless they are old, believing that they will become bi-coloured like the snake, and therefore unattractive to men.

Snakes are stretched out and held as they cook, to prevent the contracting muscles from coiling the snake up. Snake eggs are an Aboriginal favourite (I find them floury and tasteless).

I caught a water python for Gunwinggu women 18 months before my young son, Rowan, was born. His grandmother said that the snake's spirit had entered his body meaning that he and Water Python are related, ie this snake is his 'dreaming' (as is Olive Python below). This relationship means that he must look after the welfare of these snakes and is not allowed to hurt them. He cannot eat this snake even when it is killed by someone else as this would be considered cannabilism.

Similar Species: Olive Python *(Liasis olivaceus).* This snake is olive-brown and cream below, not yellow or apricot. It only has one loreal scale, like Water Python but it has smaller scales and thus a higher mid-body scale row count (55-80). It grows much longer than Water Python, having a maximum length of about 4 m. One 3m.

female dissected on the old Stuart Highway had a fully grown male bandicoot in her gut. Can be found in same area as Water Python, eg Fogg Dam. Common.

Olive Python is found across northern Australia except for eastern Queensland. Top End records from Darwin, and Cutta Cutta Caves, Fogg Dam, Manton Dam and Umbrawarra Gorge Reserves, and most Top End national parks. Other records from Adelaide River township, Douglas River, Fountainhead, the old Stuart Highway, Pine Creek, Arnhem Highway and the length of the Kakadu Highway.

Once a 4 m. Olive Python joined some friends and me as we were having a sunset snack beside the Stuart Highway near Howard Springs. The children scattered and one rather large adult managed to land on the bonnet of the car in a single leap. Knowing that it was harmless, I didn't move, and as it passed I stroked it. It gave me what appeared to be a haughty glance, as if to put me in my place, and then moved unhurriedly off.

Because Olive Python looks very much like Water Python, it too is regarded as Rowan's 'dreaming'. He takes this relationship seriously. We once found an Olive Python which was too badly injured to survive and I killed it. Rowan who had not cried at the death of his puppy some months before, broke his heart over the death of this snake that he regarded as a close relative.

Hilary Thompson has seen small birds including White-gaped Honeyeater, pursuing an Olive Python in Kakadu. I once witnessed the unedifying sight of a tour bus swerving to the wrong side of the Arnhem Highway one night to run over a large Olive Python, hardly sustainable tourism. Gunwinggu people call this snake Mandjuduk .

OENPELLI PYTHON *(Morelia oenpelliensis)*

Description: A large to very large python, pale brown with dark brown blotches and paler underneath. Irridescent by torchlight. May grow larger than 4m. It may be unwise to attempt to count the scales of large specimens (but for those who are interested, or game enough, there are 70 at mid-body).

Habits: A rock-dwelling python only 'described' in 1977 although long known to Aboriginal people. My adopted Gunwinggu son says that it eats Black Wallaroo, other mammals, and birds. He also says that it is 'very dangerous for children', that people have memories of this snake occasionally taking them. Oenpelli Python is easiest to see in the wet season when it comes down from the rocks, according to Gunwinggu people.

Where found: Restricted to Western Arnhemland including Kakadu National Park.

Notes: Gunwinggu people call Oenpelli Python Ngarawan. However my daughter, Amber, is named Garlinawa, an Aborginal name meaning 'rock-snake, big one'. This snake according to elders, grows to 'fifty feet long'. The older women say that men are terrified of this snake, calling it 'mother' as a sign of respect. As I've said to Amber, this cannot be a bad name to have. 'Garlinawa' may be another name for Oenpelli Python, or there may be another large snake waiting out there to be 'discovered'.

An Oenpelli Python became egg-bound in 1984 and was sent to Queensland for a Caesarian which was successfully carried out. The clutch size was ten eggs. Such an operation is carried out by cutting between the scales of a snake (I covered this story for a local radio station and nearly won an Australia-wide award for originality!).

FILE SNAKES
Family: Achrochordidae

This family contains three species, which occur from India and SE Asia to Australia. They are all aquatic with small keeled scales, reminiscent of a rasp, which are continuous around the body -

there are no enlarged ventral scales. File Snakes are non-venomous and usually placid, but large specimens are capable of delivering a nasty bite. The hides are used commercially in SE Asia.

File Snake

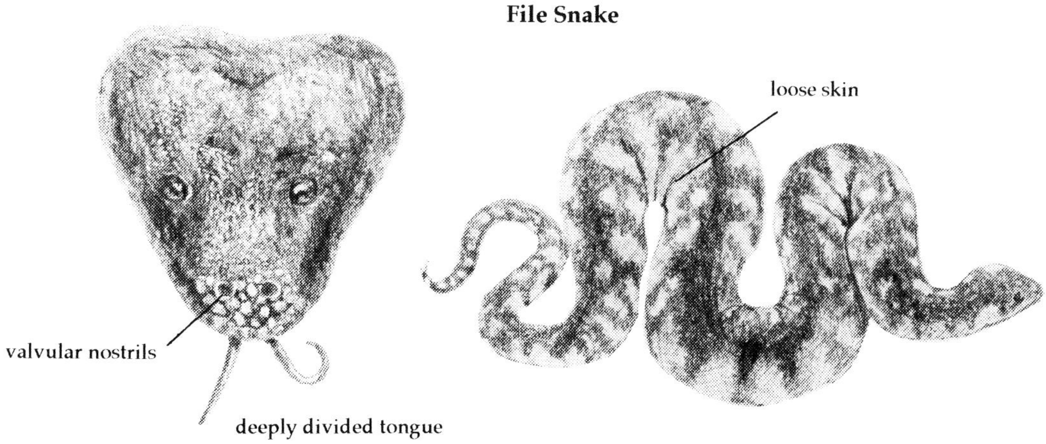

ARAFURA FILE SNAKE *(Acrochordus arafurae)*

Description: A dark grey/brown acquatic snake with a very loose, baggy skin, indistinctly patterned with granular scales. They have a very deeply-forked tongue and valvular nostrils. Average length about 1.5 m and females are bigger than males. These snakes are virtually helpless on land. Relatively common to abundant in suitable habitat.

Habit: File Snakes ambush fish underwater, from the roots of pandanus and other trees. Males sometimes attempt to mates en masse with a single female. Females give birth late in the wet season to as many as seventy live young. Large numbers of this species of file snake can sometimes be seen as floodplains dry up later in the year.

Where found: In fresh water and sometimes among flooded grass on the edges of floodplains in the Top End, northern Queensland, and southern New Guinea. Top End records from Fogg Dam and Howard Springs Reserves, Litchfield, Kakadu and Nitmiluk National Parks, and Mary River and Arnhemland wetlands.

Notes: Aboriginal women fish for file snakes, feeling with either a stick or their toes. They are apparently good to eat although their appealing face (see picture) has always put me off when I've been offered them for dinner. The women put its head in their mouth and pull on the body to kill it. Gunwinggu people call this snake Gadjaba.

Similar species: Little File Snake ***(Acrochordus granulatus)*** is smaller and banded. It is basically a marine/estuarine snake.

SOLID TOOTHED/REAR-FANGED SNAKES
Family: Colubridae

This family of snakes dominates all continents except Australia (and of course Antarctica). There are three sub-families here, Boiginae, Homolopsinae and Colubrinae. Members have large scales and usually, although not always, a loreal scale . Many are harmless, others have fangs in the rear of their mouths and are slightly venomous. Although some species of colubrid are found down the east coast of Australia, they are not found in the extreme south.

BROWN TREE SNAKE/NIGHT TIGER *(Boiga irregularis)*

Description: Rust-brown and white banded, nocturnal snake. It is rear-fanged but not dangerous. Scales 19-23 rows at midbody. Average length 1.5 m. Has a cat-like eye with a vertical pupil, and one loreal scale. Common.

Habits: This snake eats eggs and small vertebrates such as birds and bats. Emits unpleasant-smelling fluid from anal glands if handled roughly. It is an egg-layer.

Where found: Inhabits adits and caves, monsoon vine thickets and monsoon forest, and is found right across northern Australia and down the east coast, and also in New Guinea and Indonesia. Top End records from Cutta Cutta Caves, Fogg Dam and Umbrawarra Gorge Reserves, and Gurig, Kakadu, Keep River and Nitmiluk National Parks. Records from Mary River and Borroloola regions. Several were noted at night around the openings of mine shafts at Mt. Todd. I have found them dead on Top End roads. The genus is found in Africa, SE Asia, and West Pacific islands.

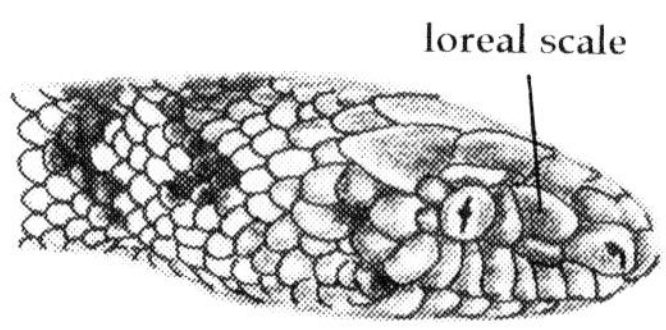

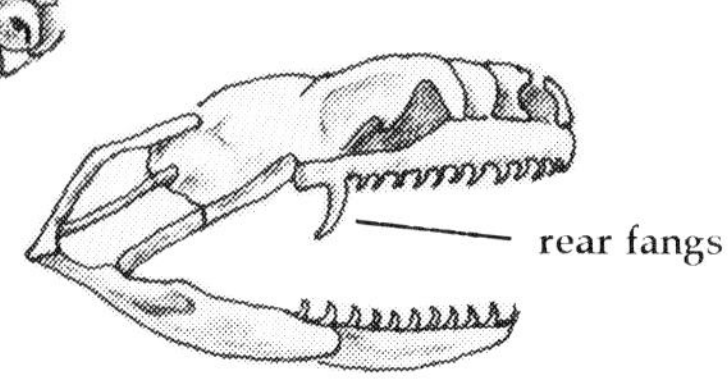

Brown Tree Snake skull

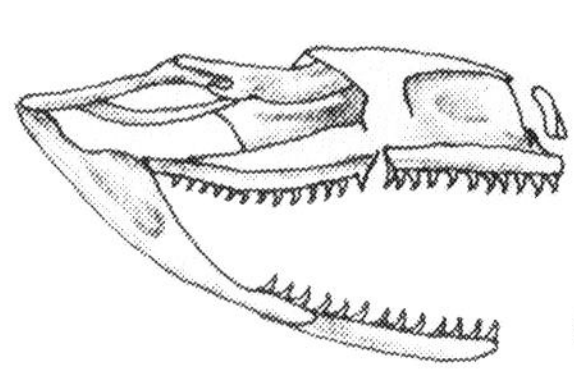

Common Tree Snake skull

COMMON TREE SNAKE *(Dendrelaphis punctulatus)*

Description: A harmless, slender snake with a blue-grey head, a grey/green back, and butter-yellow sides, belly and tail. Juveniles may be brown. (Adults may be different colours, eg blue). It has a large eye, one loreal scale and very large scales, 13 rows at mid-body. Grows to 2 m. Common.

Habits: This fast-moving snake frequents pandanus, paperbarks hunting for frogs, birds etc. It has benn noted hunting both during day and night. Like Brown Tree Snake it discharges a foul-smelling fluid when frightened. It lays up to twelve eggs per clutch.

Where found: Common Tree Snake is found across northern Australia and down the east coast, and also in New Guinea. Top End records from Cutta Cutta Caves, Fogg Dam, Manton Dam and Umbrawarra Gorge Reserves and most Top End national parks. Records also from Darwin, Bathurst and Melville Islands, Fountainhead, Douglas River, Katherine, Marrakai Plains, Mary River and Mt Bundey, Port Keats and Borroloola region.

Notes: Once on the Victoria Highway I stopped by a Common Tree Snake sunbaking on the road. It slid across to the car, raised the upper part of its body and crawled into my car. I saw a large specimen attacked and then eaten by a big goanna at Douglas River.

Similar species: No other snake is similarly coloured like the adult Common Tree Snake. However I have noted juveniles that superficially resemble Marbled-headed Whip Snake, *Demansia olivacea*.

WHITE-BELLIED MANGROVE SNAKE *(Fordonia leucobalia)*

Description: A beautiful, inoffensive, heavily patterned snake found in the mangroves. Colours and pattern vary - some are just black and white, or orange with grey or black spots or blotches. This species is rear-fanged and slightly venomous. 23-29 scales at mid-body. Unlike most other colubrids, White-bellied Mangrove Snake lacks a loreal scale. Not commonly seen.

Habits: This species feeds on crustacea, immobilising crabs by biting off their legs and claws, and fish and can sometimes be seen crawling into crab holes in the mud. Gives birth to live young in the mid-dry season.

Where found: In mangroves around the coast of northern Australia, and alsoin SE Asia, the Philippines and New Guinea. Record from Sanderson sewerage ponds and from Gurig and Kakadu National Parks, Port Keats and the Borroloola region.

Notes: One found near the Sanderson sewerage ponds was investigating a crab hole. When removed it showed no inclination to bite although handled by several people. There is only one species found within this genus.

White-bellied Mangrove Snake

SLATY-GREY SNAKE *(Stegonotus cucullatus)*

Description: A lovely, uniform grey snake with a cream belly and sides tinged with dusky-pink. It has 17 rows of mid-body scales and one loreal scale. A non-venomous snake that pretends it is anything but, biting savagely when handled and releasing a foul-smelling odour from its anal glands. 1m. Relatively common.

Habit: This snake is semi-aquatic and nocturnal and feeds on frogs, and other reptiles. It lays eggs.

Where found: In wetlands, paperbark and monsoon forest, and near creeks in the Top End, northern Queensland and New Guinea. One frequented a 44 gallon drum of water next to the Polishuk's house at Bamboo Creek, Daly River . Recorded from Fogg Dam Reserve, and Gurig and Kakadu National Parks. Also found on Marrakai Plains and Mary River wetlands. Genus occurs in south east Asia and the Philippines.

KEELBACK *(Tropidonophis mairii)*

Description: A harmless nocturnal grey / brown semi-aquatic snake with keeled scales, often spotted with black. It has 15-17 rows of scales present at mid-body, and one loreal scale. Grow to 1 m. but is usually much smaller. This snake has the un-snakelike ability to drop its tail when handled roughly. Common.

Habits: This snake feeds on frogs and tadpoles. It breeds most of the year in the Top End, and like most colubrids, is an egg-layer.

Where found: Found in the wetter parts of northern Australia and down the east coast to northern New South Wales, and in New Guinea. Top End records from Fogg Dam, sometimes in large numbers in the late dry season, Holmes Jungle (in Darwin) and Manton Dam Reserves, and Gurig, Kakadu and Nitmiluk National Parks. Other records from the Fergusson River, Marrakai Plains, Mary River, Pine Creek / Katherine and Borroloola regions.

Similar species: There is a similar-looking snake down the east coast of Australia - **Rough-scaled Snake** *(Tropidechis carinatus)*. It is sometimes found with Keelbacks. Rough-scaled Snake is an elapid, grows to about one metre and is DANGEROUS. Rough-scaled Snake is not found in the Top End.

FRONT-FANGED VENOMOUS LAND SNAKES
Family: Elapidae

Members of this family predominate in Australia. All are venomous, though not all are dangerous to human beings.

The venom of Australian elapids, is mainly neurotoxic, although there are exceptions. Whip Snakes have haemotoxic components (breaks down blood cells) and King Browns have myotoxic venom (breaks down muscle tissue). Elapids are large-scaled (although not as large as some colubrids); all in the Top End have less than 23 scales per mid-body row. No Australian elapid has a loreal scale. Many lay eggs, being either oviparous, or oviviparous (that is, the eggs hatch inside the mum, and the young are born minus egg). Others are viviparous, having a placenta.

Elapids have a fang and several teeth on the maxillary bone. Vipers only have one large fang on this bone.

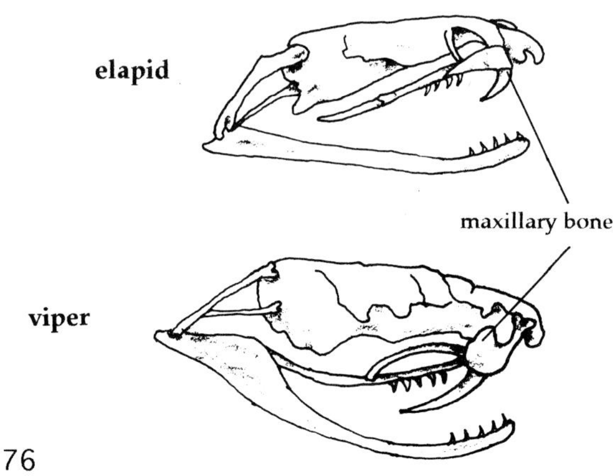

There are about fifteen species of elapid in the Top End compared with roughly 33 species in south-eastern and south-western Australia.

NORTHERN DEATH ADDER *(Acanthophis praelongus)*

Description: A small, stout, banded viper-like snake with a diamond-shaped head, keeled scales, and a thin tail terminating in a spur. This snake, unlike most other elapids has very large fangs. It has 21-23 scale rows at mid-body. Those I see at Fogg Dam are olive in colour. However this species can be highly variable. I saw some grey and white specimens near Victoria River, possibly a different species, *A. antarcticus*. Average size 0.5 m. Common in some areas.

Habits: A rather sluggish snake that lies in wait for prey then pounces rapidly. I have seen them lying buried in leaf litter twitching their tails to attract passing animals. They feed on mammals such as rats, and birds.

Where found: Found across northern Australia and in New Guinea. There is a large population at Fogg Dam. Reserve Recorded in Kakadu, Litchfield and Nitmiluk National Parks and the Mary River and at Fountainhead. Very common in the Victoria River area where they inhabit the culverts under the bridges in considerable numbers.

Notes: This snake is **DANGEROUS.** Do not walk around in the bush without proper foot protection or put hands into leaf litter or hollow logs. This snake is very hard to spot because of its cryptic colouring. Unlike other elapids, the Death Adder can fold its fangs back along the upper jaw to a limited degree. Vipers can also fold their fangs back.

Death Adder

Marble-headed Whip Snake

Black Whip Snake

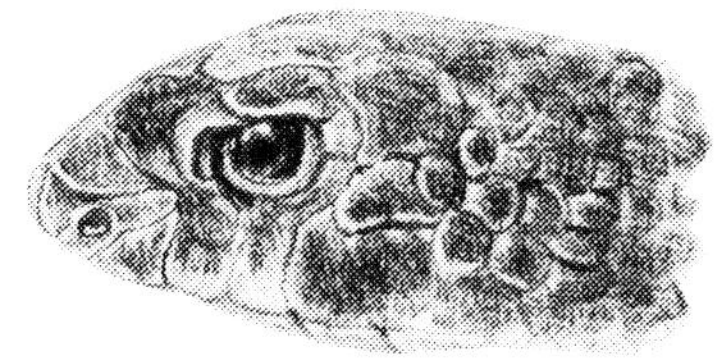

BLACK WHIP SNAKE *(Demansia atra)*

Description: A pretty, slender black snake with a pale belly and reddish undertail. Its head is unmarked. There are 15 mid-body scale rows and 70 - 95 subcaudals, all divided. A ridge over the eye-socket makes this snake, and other elapids, look rather cross. 1.5 m.

Demansia species have big eyes (see picture, previous page), a feature of diurnal snakes. Common.

Habits: A very fast snake with accurate eyesight which it uses to locate prey. It is mostly diurnal, hunting during the morning and late afternoon, but can be seen at night during hot, humid weather. A nervous snake - I watched one stalking a small rodent from a bridge on the Victoria Highway and it picked up the vibrations of a car coming along the highway a kilometre or two away, and quickly disappeared. It eats frogs and lizards as well. An egg-layer.

Where found: Found over most of northern Australia and down the east coast to the New South Wales border. Top End records from Cutta Cutta Caves, Fogg Dam and Manton Dam Reserves and Gregory, Gurig, Kakadu and Nitmiluk National Parks. Recorded all the way down the Stuart Highway to Mataranka, the Arnhem Highway at Beatrice Lagoons, Marrakai Plains, Mary River, and into Jabiru. Also down the Kakadu Highway and the Victoria and Carpentaria Highways.

Notes: Whip Snakes have a haemotoxic component in their venom. A bite causes great pain but not death, although large specimens should be considered dangerous, especially to children. Like most snakes, they are more interested in escaping than stopping to sharpen their teeth on someone. Whip snakes look very much like the relatively harmless colubrids found on other continents.

Related species: ***Demansia papuensis*** is very similar. However it is larger (max. length about 2 m.), has a spotted head, lacks the reddish undertail (it is white or blueish instead), and has a longer tail. It does not seem to be as common in the Top End as *D. atra*. It is found across most of northern Australia in savanna woodland, and in New Guinea. Top End records from Blackmore River, Mary River and Umbrawarra Gorge Reserve. Also Kakadu, Keep River, Litchfield and Nitmiluk National Parks.

Marble-headed Whip Snake *(Demansia olivacea)* is smaller and light-coloured, being olive-brown or grey above, with a distinctly patterned head. It is found in the Top End and down the Western Australian coast. Top End records from Gurig, Kakadu, Litchfield and Nitmiluk National Parks. This species does not seem to be as common as *D. atra* in the Top End. (See picture previous page).

MOON SNAKE/ORANGE-NAPED SNAKE *(Furina ornata)*

Description: This tiny, inoffensive orange snake is found at night, usually in rocky areas. It has a reddish collar and a dark brown head which looks quite long-nosed for an elapid. Each scale is dark-edged, with a netted appearance. Those seen by the author have been around 30 cm. long. Venomous, like all elapids, but too small to hurt humans of any size.

Habits: This snake can be very active during the build-up months. It hides under rocks and logs during the day. It eats lizards and occasionally invertebrates. An egg-layer.

Where found: Widespread through northern and central Australia. In the Top End it is often found in stony hills. Top End records from Darwin, Cutta Cutta Caves and Fogg Dam Reserves, and most Top end national parks. Also in the Borroloola and Pine Creek regions.

Similar species: Half-girdled Snake *(Simoselaps semifasciatus)* A tiny inoffensive black and brown-banded burrowing snake growing to about 20 cm. It has white/cream underparts and a cute, upturned nose with a hard scale on the tip. Can sometimes be found under rocks or seen above ground at night in the late dry season/early wet season. Found across most of Australia. Top End records from Kakadu, Keep River and Nitmiluk National Park. (See picture, page 81).

Blind Snakes These harmless burrowing snakes belong to a different family, Typholophidae. Various species found all over the Top End. See page 69).

TAIPAN *(Oxyuranus scutellatus)*

Description: This large, slender **DANGEROUS** snake is light to dark brown with a pale face and an elongated head. Scales on the neck and down the spine are keeled. Mid-body scales 21-23 rows. Unlike most elapids this species has large fangs followed by only one or two teeth. It grows to three metres. This species is uncommon in the Top End. (See picture next page).

Habits: Can be found during night or day (probably depending on temperature) but is usually diurnal. Unlike most other elapids the Taipan feeds almost exclusively on warm-blooded prey, eg marsupials, rodents and occasionally birds. It lays up to 25 eggs.

Where found: It is found across Northern Australia and down the east coast. and in the New Guinea lowlands. There are few Top End records, mainly from heavily grassed areas and monsoon forest, where there may be large populations of rats. Recorded from Kakadu and Litchfield National Parks, and from Daly River (Nancy Polishuk) and Melville Island (Paul Horner).

Notes: Nancy Polishuk tells the story of an encounter with a taipan in her book, *Life on the Daly River*. Her children chanced upon a taipan taking eggs in the flowlyard The snake moved towards them but luckily was stopped by the wire netting where Nancy shot it. She notes that King Browns tended to retreat whereas Taipans either stood their ground or attacked.

This snake will deliver several bites in succession, holding on and chewing to release venom. It is closely related to the Fierce or Small-scaled Snake, *O. microlepidotus* which is considered to have the most potent venom in the world. This latter species is not found in the Top End. The venom has both neurotoxic and haemotoxic components. Gunwinggu people call this snake Yirrbudbud. My Gunwinggu son says it is a 'cheeky one', meaning that it is very dangerous.

KING BROWN/MULGA SNAKE *(Pseudechis australis)*

Description: Another large DANGEROUS snake. The head is big and blunt and almost triangular with a protruding eye-brow. This snake is light brown, and sometimes the scales are outlined in a darker colour to give a netted effect. Mid-body scale count 17 rows. I measured one dead specimen at 2.75 m. Another that personally introduced itself by sliding up to my feet, was even longer. (Picture next page). Nancy Polishuk, writes that King Browns were common when the river was in flood - she saw 'at least one a day'.

Habits: This snake is usually diurnal although I have seen them at night in hot, humid weather. In late dry season they tend to move around in early morning and on cool days I have seen them at midday. It feeds on mammals, frogs, and other reptiles - I found a Half-Girdled Snake in the stomach of a King Brown on the Kakadu Highway. Big specimens are common. I counted three roadkills, all over 2 m in length, in August, 1993, near the township of Adelaide River.

Where found: This snake is found over most of Australia. There are records throughout the Top End, including Darwin. Also recorded from Cutta Cutta Caves, Fogg Dam, Katherine Low Level Crossing, Manton Dam and Umbrawarra Gorge Reserves, all Top end national parks. Other records from Mt Wells and the Fergusson, Douglas and Mary Rivers and the Borroloola and Pine Creek regions. Often found in caves and adits.

Notes: The author has found this snake to be rather inoffensive if left alone. Nancy Polishuk writes in *Life on the Daly River* that she found them 'less belligerent' than the Taipan. However the King Brown has a reputation for being very nasty when cornered or trodden on. Venom is myotoxic and bites can result in gangrene, if not death. It has the unpleasant habit of holding on and chewing, thereby injecting more venom. Even herpetologists have been known to mistake this snake for a python in the dark so be careful. Gunwinggu people call it Datbar.

WESTERN BROWN SNAKE *(Pseudonaja nuchalis)*

Description: A very pretty but **DANGEROUS** yellow-brown, tan or olive-brown snake often with darker blotches. Each scale may have darker edges giving a netted effect. There is often a V or W shaped mark on the nape, or the head and neck may be black. Sometimes the snout and lower jaw are pale, a little like the Taipan. Scales number 17 -19 rows at mid-body. This snake grows to 1.5 m. A common species.

Habits: Diurnal and nocturnal (seems to depend on temperature). I have seen this species raise their head and body a metre in the air, as if to look around. They can also rear up like this when injured or cornered and it is not a nice feeling to get out of your car to examine an injured Western Brown and find it attacking you at eye-brow level. This snake feeds on small mammals and frogs. Oviparous, laying up to 20 eggs.

Where found: This snake is found over most of Australia. Common in Top End in open forest and woodland. People have been bitten in Darwin. Records from Cutta Cutta Caves, Fogg Dam, Manton Dam and Umbrawarra Gorge Reserves, and most Top End national parks. Other records from Fountainhead (where Peter Nicholls and I found one with its head trapped in a discarded beer can), Mary River, the Borroloola region and along all highways. Six passed me at sunset, in the space of five minutes, while I was sitting counting bats beside a mineshaft at Mount Todd.

Taipan

King Brown

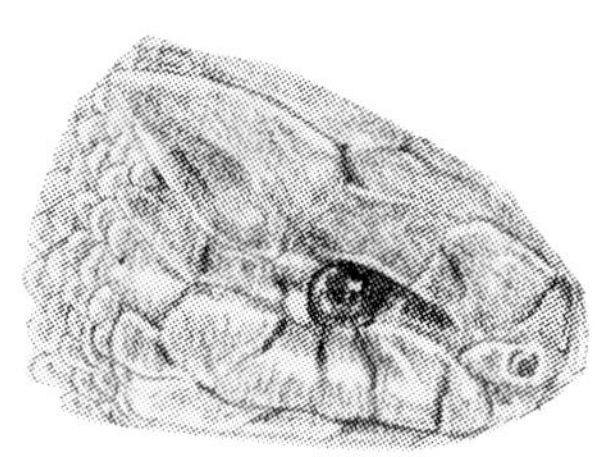

Western Brown Snake

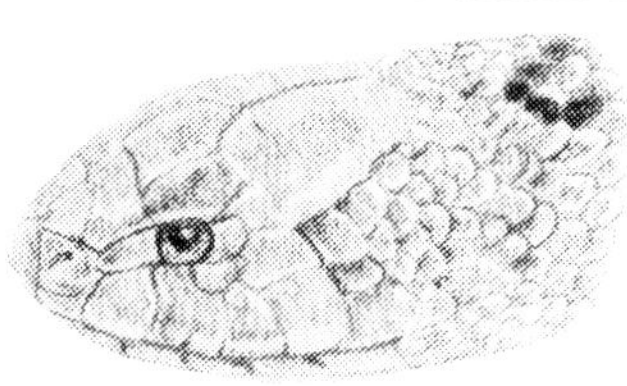

NORTHERN BANDY BANDY*(Vermicella multifasciata)*

Description: An inoffensive nocturnal black and white banded snake that grows to about .75 m. One of two species of black and white banded snakes in Australia. Both belong to this genus. This snake is venomous but is not considered dangerous to human beings. Uncommon. (Picture, next page).

Habits: Prey consists of insects, blind snakes and lizards. Has a peculiar defensive position in which it throws its body into loops. An egg-layer.

Where found: Under rocks or out in the open on warm nights. This species is confined to the Top End and north eastern Western Australia. Top End records from Cutta Cutta Caves Reserve, Kakadu and Nitmiluk National Parks and Mt Todd.

Similar species: Half-girdled Snake *(Simoselaps semifasciatus)* See page 78).

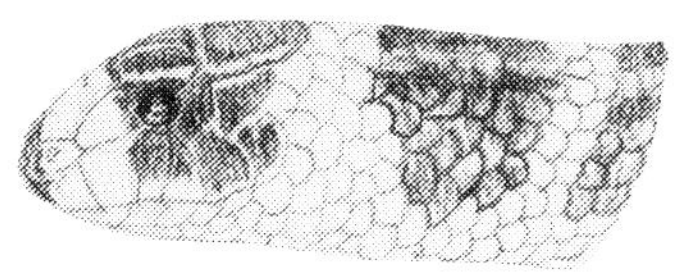

Northern Bandy Bandy

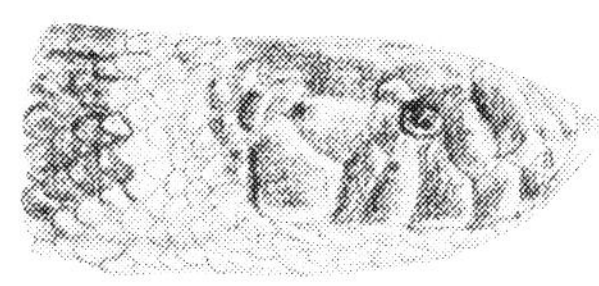

Half-girdled Snake

FROGS
Class: Amphibia
Order: Salientia

Amphibians (meaning 'two lives') refers to the fact that these animals spend part of their lives as larvae (tadpoles) and then metamorphise into adults. The word 'tadpole' is Middle English, and means 'toadhead'.

Traditionally the term 'frog' is reserved for those tail-less amphibians with smooth, moist skins. Those terrestrial amphibians with warty skins are called 'toads'. However many frogs also have warty skins and terrestrial habits.

Amphibians were the first vertebrates to colonise the land, about 350 million years ago. However they were restricted by their inability to live without water because they had permeable skins which aided in respiration* and therefore had to be moist, and their eggs, like those of fish, lacked waterproof shells and had to be laid in, or near, water. Many ancient amphibians were giants, some growing to three or four metres. *Paracyclotosaurus davidi,* which inhabited several parts of Australia, measured about two metres, excluding tail. These early animals resembled the newts and salamanders.

Fertilisation in mammals, reptiles (except for the Tuatara of New Zealand) and some amphibians, eg newts, salamanders and another group called the caecilians, is internal. Newts and salamanders just bring their cloacas together. The male caecilian everts his cloaca and uses it like a penis. Fertilisation is external in frogs and toads, meaning it occurs outside the female body.

Frog mating is called 'amplexus'. The smaller male clings to the female's back, then brings his hindfeet inwards to touch her cloaca which she raises to squirt out the eggs, like a miniature volcano. He then fertilizes them. Eggs are deposited in different places, in water or on land. Some frogs both in Australia and overseas, carry them about on or even in the body until they hatch. (See Myobatrachidae below).

Males and females look similar except males are bigger. Most males have special pads on their fingers called 'nuptial pads' which help them to hang onto slippery females. Likewise some female frogs have widened digits which enable them to make foam nests for their eggs. A frog's hand has four fingers and the foot has five toes.

*their lungs, like those of existing amphibians, were incapable of providing all their oxygen intake. In fact some, like American salamanders today, probably did not have lungs, and simply relied on their moist skin and the lining of their mouth. Unlike higher vertebrates amphibians which do have lungs, have to gulp air actively.

Frogs are fascinating creatures. For instance hibernating frogs in cold countries contain glycerol (used in anti-freeze) in their body fluids.

During the day frogs' eyes have pupils which appear as a slit, much like those of a cat. During the night however, these expand and the eye looks black.

Frogs seem ill-equipped to handle the nine months or so of drought that occurs in the Top End every year. However they do, to reappear in large numbers during the wet season. During the dry season some species seek shelter in the cracking clays of the floodplains. Others survive within hollow trees or by burrowing into mud, which later dries. In the wet season they breed rapidly - tadpoles congregate by the millions in the floodplains at the edge of flooding billabongs and creeks.

Lots of animals prey on frogs, including snakes, mammals such as Northern Quolls, and water-birds. Rufous Night-herons line up on the cause-way at Fogg Dam at night to catch frogs.

All continents (except Antarctica and Australia) have endemic toads, newts and salamanders. There are four families of frogs native to Australia. Toads have been recently introduced-the Cane Toad *(Bufo marinus)* of the family Bufonidae. (See section of Introduced Animals). There are over 3000 species of frogs in the world, about 200 species in Australia, and about 35 in the Top End.

Gunwinggu people call all frogs Gobolbuk or Djarti.

TREE FROGS
Family Hylidae

Most of these frogs have features which make them efficient climbers. For instance most adhesive have discs on their toes (apart from one genus, *Cyclorana* species), allowing them to cling to objects. They also press their abdomen against a flat surface creating suction.

Hylid frogs are found in other areas of the world, eg Central and South America. However they may not be as closely related as once thought, and there are moves afoot to put the Australian hylids into a separate family.

BURROWING FROG *(Cyclorana australis)*

Description: A large, solitary burrowing frog often found on roads during the more humid times of the year It is usually grey/brown above and white below, but sometimes has emerald green patches. Most of those noted by the author during the wet season have been green, probably juveniles. There is a broad dark stripe through the eye. The head is big with a blunt snout and there is a pronounced dorso-lateral fold (the fold of skin that runs parallel with the spine and is situated halfway between the spine and the side). Unlike most members of Hylidae it has no discs on its digits. 10 cm. Common. (See picture next page).

Habits: *Cyclorana* species burrow backwards into the soil with the use of special tubercules on their back feet.

Where found: This frog is found in much of the Northern Territory and northwest Western Australia. It is commonl all over the Top End in a range of habitats including savanna woodland and floodplains. Recorded from most national parks, and Fogg Dam and Manton Dam Reserves. Also Mary River wetlands /Marrakai Plains, in the Fountainhead/Pine Creek/Katherine and Borroloola regions, and down the Victoria Highway to Timber Creek .

Notes: This frog is related to *C. platycephala,* the water-holding frog traditionally used by desert Aboriginal people during droughts. As surface waters evaporate this frog fills its bladder with water, burrows into the ground and forms a transparent cocoon of sloughed dead skin around its body which helps prevent water loss. One is known to have survived in captivity for sixteen years suggesting that *C. australis* may have a similar life span.

Similar species: Green Tree Frog (see *Litoria*) can grow even bigger. However unlike Burrowing Frog it has large discs on its digits and no facial stripe.

Cane Toad ***(Bufo marinus)*** is larger (15 cm) and wartier with huge parotid glands. It lacks discs like Burrowing Frog but also lacks a facial stripe. It is an introduced member of a fifth family, Bufonidae. All occurrences in the Top End should be reported to the Northern Territory Conservation Commission. (See Introduced Animals, page 89).

Litoria wotjumulensis grows to 7-8 cm and looks somewhat like a small burrowing frog. However it is slimmer with a pointy nose and lacks a dorsal-lateral fold. It also has tiny discs which Burrowing Frog lacks. See *Litoria.*

Burrowing Frog

Green Tree Frog

Cane Toad

NORTHERN DWARF TREE FROG *(Litoria bicolor)*

Description: a small green or brown bi-coloured frog often found resting on pandanus fronds, reed stems, or on the road at night near floodplains. It is white below, the dorsal and ventral colours being divided by a broad dark band along the side. 3 cm. Common.

Where found: Found across northern Australia and throughout Top End. Records from Darwin. Also Fogg Dam Reserve and Gurig, Kakadu, Keep River and Nitmiluk National Parks. Also Mary River wetlands, Fountainhead, Pine Creek/Katherine and Borroloola regions.

Similar species: Could be confused with immature Green Tree Frog. However the latter has no dorso-lateral stripe.

GREEN TREE FROG *(Litoria caerulea)*

Description: A large green tree frog often found around houses in Darwin. It is white underneath with large discs on its digits. The backs of the thighs are bright orange-yellow with no markings. Females are bigger than males, 12 cm. Juveniles are quite small about 3-4 cm. in length. Common.

Habits: These frogs call from hollow trees and downpipes making a sound reminiscent of a digeridoo. I found the croaking of Green Tree Frogs evocative, until seven moved in downstairs. Since then my love for them has diminished along with my sleep.

Where found: Found across northern Australia and down the east coast. Common in Darwin where they may be found living in lavatories and letterboxes. Records from most of the major national parks, and Fogg Dam and Manton Dam Reserves. Also Mary River wetlands, Fountainhead/Pine Creek and Borroloola regions.

Notes: Green tree frogs contain a substance caerulein, which is sometimes used to restore intestinal function after surgery.

SAXICOLINE TREE FROG *(Litoria coplandi)*

Description: A small (4 cm), relatively plump, mottled light and dark brown frog with a round head and no facial stripe. Its skin is smooth or fine-grained, and it has moderate discs on digits. Relatively common in suitable habitat.

Where found: On rocky outcrops and escarpment in the Top End and northwestern Western Australia. Records from most national national parks and Douglas River, Pine Creek and Borroloola regions.

Similar species: *Litoria inermis* is also mottly brown and lacks a facial stripe, but is slightly smaller and wartier, with a pointy nose and very small discs. It is found in a variety of habitats eg floodplains, savanna woodlands and monsoon forest, across northern Australia. Records from Fogg Dam and Manton Dam Reserves and in most national parks. Other records from Douglas and Daly Rivers, and Mary River, Fountainhead, Pine Creek and Borroloola regions. Tadpoles have a white patch on the lower abdomen. Relatively common.

Saxicoline Tree Frog

Litoria inermis

Dahl's Frog

Litoria nasuta

DAHL'S FROG *(Litoria dahli)*

Description: A medium-sized (7 cm) slender brown/green frog with a pointed snout and a pale green stripe down the middle of its back. It has small discs on its digits. Common.

Where found: On floodplains and swamps across the Top End and in northwest Queensland. Sometimes during the wet season they are in large numbers on the causeway at Fogg Dam and on the Arnhem Highway near Beatrice Hill. Records from Kakadu National Park and Mary River.

Similar species: ***Litoria nasuta*** This name means long or large-nosed. Smaller than *L. dahlii* (5 cm) it has a similar shape but its snout is much longer. It has warty skin, often with a striking pattern like a footpath with crazy paving! There is a dark stripe through the eye and the digits bear small discs. Very jumpy and consequently hard to get a look at. Inhabits floodplains and edges of waterholes across northern Australia and down the east coast. Records from Fogg Dam, Manton Dam and Butterfly Gorge Reserves and most Top End national parks. Also found in Mary River wetlands and Borroloola region. Common.

LITORIA PALLIDA

Description: A small brown or grey-brown frog with darker mottling and very small discs. It has a thin black stripe from snout to forelimb, which is bordered by a paler stripe. There is a dark blotch between the eyes. 3.5 cm.

Where found: Found in open forest and floodplains. across northern Australia. Records from Manton Dam and Umbrawarra Gorge Reserves and Keep River and Nitmiluk National Parks. Also from Mary River wetlands/Mt Bundey area and in Fountainhead/Pine Creek and Borroloola regions.

Notes: Remains found in Ghost Bats' roosts.

Litoria tornieri is similar in size, shape and colour to *L. pallida* but it is not blotchy. It has very small discs and smooth skin. There is a dark stripe running from snout to forelimb where it breaks into blotches, and also a dark stripe down the front of the hind leg. It is found in in open forest and floodplains across the Top End and northern Western Australia. Records from Kakadu, Gurig, Litchfield and Nitmiluk National Parks. Also from Douglas River, Marrakai Plains and Pine Creek and Borroloola regions. Hilary Thompson and I found lots of them in shower blocks at a lodge near Douglas River in about 1989. In May 1993 there were hundreds in the grass along the Arnhem Highway near the South Alligator River during daylight. Common.

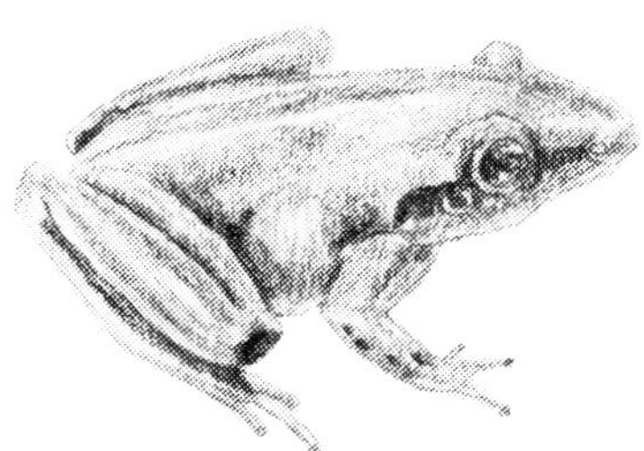

Litoria tornieri

Litoria pallida

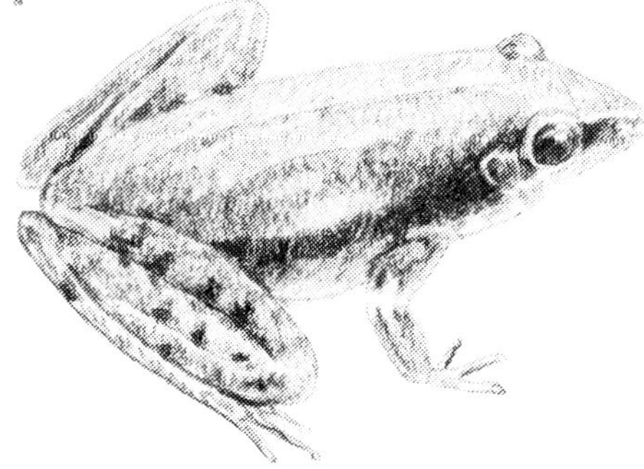

Litoria wotjulumensis

Red-eyed Tree Frog

Desert Tree Frog

L. wotjulumensis is larger (4-7 cm) than the others with a more extensive dark stripe (snout to flank) and a large head and pointy snout. It is pale to mid-brown with darker speckles, and has small discs. As with *L. pallida* there is a pale stripe running along the edge of the darker stripe. This frog is named after the Aboriginal settlement of Wotjulum in Western Australia. It is found in open forest and along streams across the Top End and into northwestern Western Australia. Records from most Top End national parks and at Mantom Dam and Umbrawarra Gorge Reserves. Other records from Douglas River, Fountainhead, Pine Creek and Borroloola regions. (picture previous page).

RED-EYED TREE FROG *(Litoria rothi)*

Description: A cream or brown frog, plain or patterned. It has large discs on digits and a narrow dark line from behind the eye to the axilla. The upper half of the eye is red, but this may be hard to see. Has a maniacal cackle or bleat. 5-6 cm. Common. (Picture previous page).

Where found: Found across northern Australia and into New Guinea. In Darwin it can be heard (or seen) at the City Golf Course, and in the township of Adelaide River at the War Cemetery. There are records from Butterfly Gorge, Fogg Dam and Manton Dam Reserves and just about every national park. It will serenade you at the units at Yellow Waters in Kakadu. Other records from Fog Bay, Mary River,and the Pine Creek and Borroloola regions.

Notes: Frogs' calls often have a ventriloquial quality. However there is a technique called triangulation that you can use to track down these elusive blighters. Take three people, and get them to move, from different directions, towards the area from whence the call seems to come. Then use a torch to locate the frog. Remember, if you look for frogs in wetlands or near streams, that estaurine crocodiles, may be in the same vicinity.

DESERT TREE FROG *(Litoria rubella)*

Description: A pale to medium brown/grey frog with a small head, a thin black stripe from snout to armpit, and large discs on its digits. 3-4 cm. in length. (Picture previous page).

Where found: Found over most of Australia in a wide variety of habitats. Records from Butterfly Gorge and Fogg Dam Reserves, and Gurig, Kakadu, Litchfield and Nitmiluk National Parks. Also on the Marrakai Plains and in the Pine Creek and Borroloola regions

Similar species: Looks a little like Red-eyed Tree Frog but has much shorter hind legs.

THE GROUND OR SOUTHERN FROGS

Family: Myobatrachidae

These frogs are found only in Australia. Some members brood their young in their stomach. If the reader wants an informative, not quite straight book on this intriguing topic read Michael Tyler's delightful little book, *There's a Frog in my Throat (deleted) Stomach,* published by William Collins (1984) The two species of Gastric-brooding Frogs are not found in the Top End but in Queensland. Both are considered endangered.

CRINIA BILINGUA

Description: This tiny frog is very difficult to see. However it is included here because of its very loud, ratchet-like call. Don't bother trying to track it down unless you possess the eyesight of an eagle and the patience of Job. For the eager beavers, it is brown with a darker stripe along its back and fringes on its toes, but no discs on either digits. 2 cm. Common.

Where found: In swamps and floodplains across the northern Top End and into northern Western Australia. Records from Fogg Dam and Manton Dam Reserves, and Gurig, Kakadu, Keep River and Litchfield National Parks. Also Mary River wetlands, Marrakai Plains and Pine Creek.

Similar species: ***Litoria meiriana*** is another tiny (2 cm) dark brown frog. It is much easier to spot. than the frog mentioned above It has obvious discs on its toes and rusty patches on its sides and is found hopping around in rocky pools during the day. Paul Horner of the Northern Territory Museum, has pointed out that it is the only frog to use the surface tension of water for jumping. The name *'meiriana'* is an Aboriginal word meaning 'rockhole'. This frog is found across the Top End and in northern Western Australia. Recorded from Butterfly Gorge, Fogg Dam and Umbrawarra Gorge Reserves, and most major national parks in the Top End. Common. *Litoria* species belong to the family Hylidae. (See page 82).

Dwarf Rocket Frog *(Litoria microbelos)* An even smaller frog (1.5-2 cm). However it is also relatively easy to see. This frog is slender with obvious discs on its toes, s a darker strip from the snout through the eye to the flank, and white on the sides and underside. Dwarf Rocket Frog inhabits leaf litter in closed forest some distance from water, or can be found in swamps and near creeks. If you see what appear to be crickets hopping around in these habitats, have another look as it may be this frog. Dwarf Rocket Frog is found across coastal and sub-coastal northern Australia. Recorded from Holmes Jungle and Fogg Dam Reserves, and Gurig and Kakadu National Parks. Also records from Fog Bay and Fountainhead..

MARBLED FROG *(Limnodynastes convexiusculus)*

Description: A medium-sized (4-6 cm), plump, ground-dwelling frog. It is patterned with brown or black blotches on a paler background, and has white underparts Its digits lack discs. The female has flanges on first two fingers for making a foam nest. Call is a short musical 'plonk' made repeatedly. Common. (Picture next page).

Habits:. Makes a foam nest in small holes in damp ground for its eggs.

Where found: Inhabits edges of swamps and floodplains across sub-coastal northern Australia and New Guinea. Present in Butterfly Gorge and Fogg Dam Reserves, and all major Top End national parks. It is also found in Darwin suburbs, Mary River wetlands and Pine Creek area.

Notes: Ground-dwelling frogs like the above and *Cyclorana* species have tubercules (hard swellings) on their feet to aid in burrowing. The shape of these and their position can be used to identifydifferent species. 'Limnodynastes' means 'lord of the marsh'.

Similar species: ***Limnodynastes ornatus*** is smaller and not so strikingly patterned., with a crown-shaped pattern on its nape. Found in a range of habitats across Northern Australia and down east coast. Recorded from Parap in Darwin and Fogg Dam Reserve, and most national parks, Mary River wetlands, Fountainhead and the Pine Creek/Katherine region Discovered one on my doorstep at the Pine Creek Motel in May 1993. Common. (Picture next page).

Another medium-sized patterned frog is ***Litoria nasuta*** which belongs to the family Hylidae. It has a much pointier snout than the others and discs on its digits. See under *Litoria*.

Marbled Frog

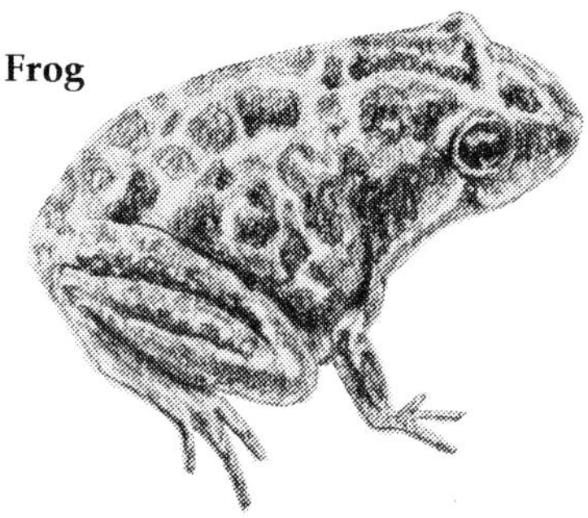

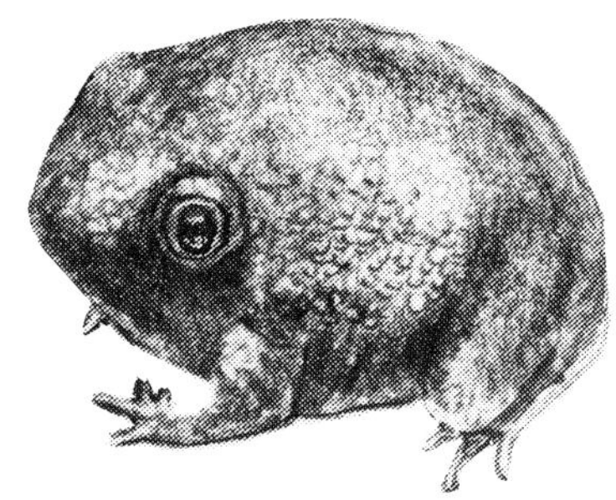

Northern Spadefoot Toad

Limnodynastes ornatus

NORTHERN SPADEFOOT TOAD/ GOLFBALL TOAD *(Notaden melanoscaphus)*

Description: A weird-looking, rotund, warty frog with short legs and long digits. The adult is dull grey/brown with darker blotches and the juvenile is brightly coloured with grey blotches outlined in yellow, and red spots. 5 cm.

Habits: This burrowing frog is only found above ground late in the buildup or in the wet season. It feeds on ants and termites. Its call is 'whoo-whoo', sounding a little like Barking Owl.

Where found: This frog is found in savanna woodland almost right across northern Australia. Records from the Darwin/Howard Springs area, Marrakai Plains, Kakadu National Park and the Borroloola region.

INTRODUCED ANIMALS

There are several species of introduced animals in the Top End. Some such as Water-buffalo, have caused major habitat damage.

Cat *(Felis catus)* Large numbers of cats were released down south in the early 1880's to tackle the rabbit problem. They have now become a problem in their own right. Both domestic and feral cats account for the deaths of several million native mammals, birds, reptiles and frogs a year.

Some town councils are attempting to control these predators with curfews, licensing and sometimes desexing of domestic cats. Unfortunately we have legitimised the keeping of cats in Aboriginal eyes and people moving onto outstations throughout Arnhemland now take cats with them both as pets and for food. I remember camping in about 1979 near the East Alligator. At dusk we were surrounded by a dozen pairs of eyes, all cats.

In Western society the rights of the individual animal and the individual human to own that animal are paramount. That animal's effect on the community or ecosystem is of minor consequence. People who believe they have a right to keep cats and that cats have a right to run free, infringe the rights of others to keep birds, and to have a diversity of wildlife in both their gardens and the bush. They also infringe the right of native wildlife to exist in harmony with their surroundings.

A tame one-winged Brown Falcon that lived in my backyard for three years, was killed by cats, Rowan, has lost both of his pet Cockatiels and Hilary his canary. Six bird species bred in my backyard last wet season and cats constantly stalked their nests.

Feral horse *(Equus caballus)* are found throughout the Top End. **Timor Ponies** are established in Gurig National Park. Both cause erosion and water pollution.

Water-buffalo *(Bubalus bubalis)* are found throughout the Top End, down as far as Mataranka. They were introduced by European settlers in the early 1900s and released when the first settlements failed.

Water-buffalo breed throughout the wet season when feed and water are abundant. Cows and their calves form maternal herds led by an older female while adult males wander alone or in loose groups. Young bulls are driven from the herd, not by the cows, but by mature bulls seeking to mate with the cows. When breeding has concluded the dominant female threatens the adult bulls, even charging them, until they depart.

These animals are normally placid. I have often scared buffalo, including one with a two metre spread of horns, away from my camp by yelling. However there are many records of people being chased up trees or gored, so never approach buffalo, especially if they have calves..

Buffalo cause much environmental damage. However they are beloved by tourguides, being easily recognisable and impressive. Also buffalo kept waterholes clear of vegetation so that waterfowl could be easily seen. Consequently many in tourism were not happy when most water-buffalo were eradicated by the Northern Territory Government in order to control tuberculosis and brucellosis for the beef industry. There are still good herds in Arnhemland.

Because of the eradication campaigns those few buffalo who escaped now feed at night. Water-buffalo are now largely domesticated in the Top End and much of their meat is exported overseas.

Banteng Cattle *(Bos javanicus)* were brought in by European settlers from Timor and are now established in Gurig National Park. They are attractive animals, with white stockings and a white 'bullseye' on their rumps.

Sambar Deer *(Cervus unicolor)*
This large brown deer is found in Gurig National Park. It was released there in the early 20 century and is now hunted by paying customers, along with Banteng Cattle. It is rarely seen, being a nocturnal and shy animal.

Pig *(Sus scrofa)* are well-established in northern Australia, probably in their millions. They grow to a prodigious size and have large litters. They cause much devastation in monsoon and paperbark forests uprooting and killing young trees and shrubs.

Cane Toad *(Bufo marinus)*
This native of Central and South America was introduced to Australia by Queensland cane-growers in 1935, in the hope that they would control native beetles that were eating sugar-cane. The president of the New South Wales Naturalist Society, Walter Froggart, tried to warn the cane-growers against introduction, and was denounced as a crank.

The toads failed miserably as beetle-controllers because they were nocturnal and the beetles were diurnal. Because they had no natural enemies (those animals that ate them died because these toads are poisonous), they multiplied prodigiously. Cane Toads have reachedBorroloola and are still moving west and north. They may reach Kakadu National Park in a decade or two if not before.

Some say that they will wipe out most small native fauna. Others say that they will have little impact. The truth is probably somewhere in between.

BONES, TRACKS, AND ANIMAL POO

Bones, tracks and animal poo* are invaluable and non-intrusive ways of identifying the vertebrate animals of an area.

*Some have already taken me to task for referring to body waste as 'poo', rather than the scientific term 'scats'. I have used the latter term because most children (and some adults) I asked, either were not familiar with this usage, or only knew that 'scat' was a term directed very loudly, in an abusive tone, at stray cats.

SKELETONS

This section is divided into Mammals and then Reptiles. Teeth, skulls and other bones with the exception of echidna which lacks teeth, are covered. Only those differences that are easily located by the layperson are highlighted. Drawings are roughly to scale.

MAMMALS

ECHIDNA

Skull
Echidnas have a bird-like skull with a long pointed muzzle. However an echidna's skull is bulbous, unlike a bird's and the nostril is located near the tip of the muzzle. Birds' nostrils are located nearer the eye-socket.

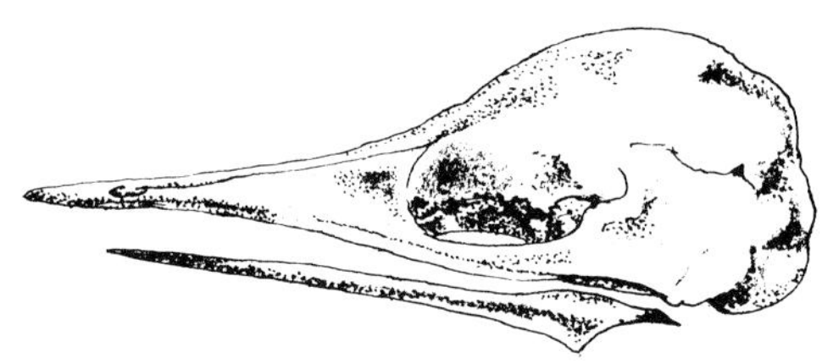

Echidna skull

Both birds and echidnas lack teeth. However, unlike birds, echidnas have a bony palate.

HERVOROUS MAMMALS

Marsupial skulls may be confused with those of other mammals. Here are ways to tell them apart.

Teeth
Kangaroo and wallabies, rodents, cattle and goats all lack canine teeth, having instead a long gap - the diastema, between the front and back teeth (the front teeth are called incisors and the back teeth pre-molars and molars).

Kangaroos and wallabies* have three pairs of upper incisors while rodent skulls have only one. Goats, buffalo and cattle lack upper incisors altogether, having instead a hard pad which the lower incisors bite against.

*The Potoroo, a type of kangaroo, has canine teeth but is not found in the Top End

cow skull

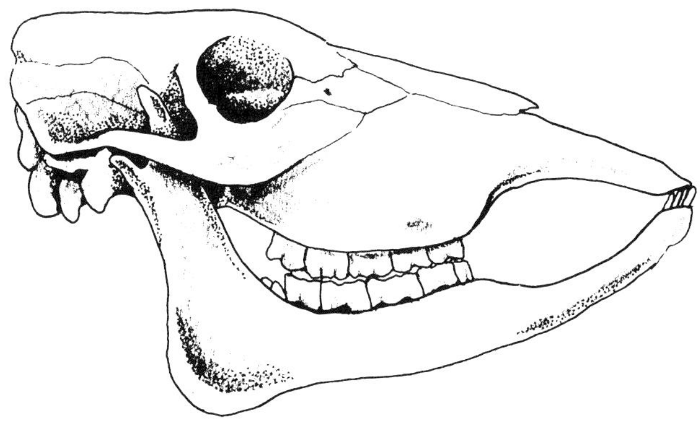

kangaroo skull

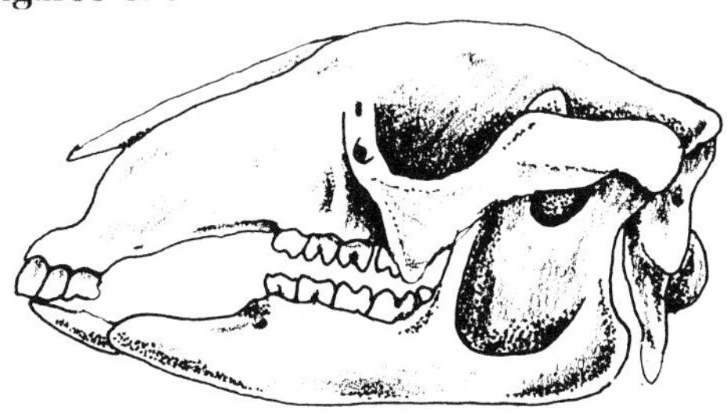

rodent skull

Horses and pigs are a little different. They also have three pairs of upper incisors. However they have three pairs of lower incisors (kangaroos have only one). These teeth are small - the kangaroo's are large and chisel-shaped. Horses usually have small canine teeth in both jaws. Pigs also have canine teeth, but these can be large, forming tusks in males.

Possums' teeth are similar in number and arrangement to those of kangaroos. However they also have canine teeth.

horse skull

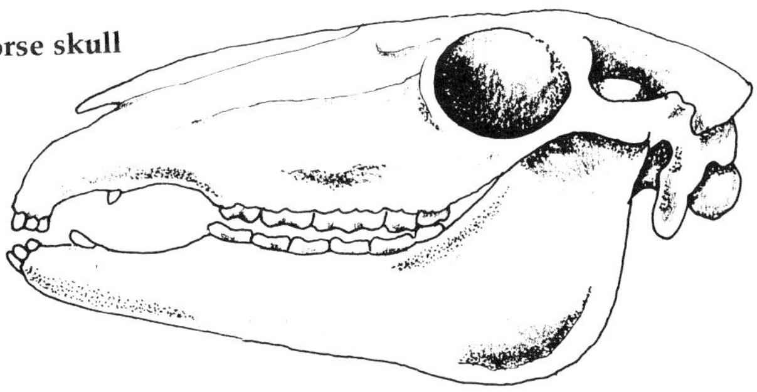

possum skull

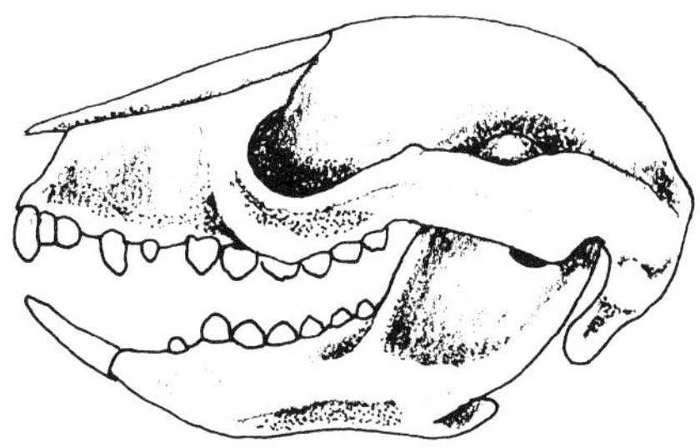

Molars difer widely between different mammals as well. Those of cattle, buffalo, goats and possums have crescent-shaped grinding surfaces, while those of kangaroos have a sharp ridge running across each molar. Pigs' molars have rounded grinding surfaces similar to humans'.

cow molars

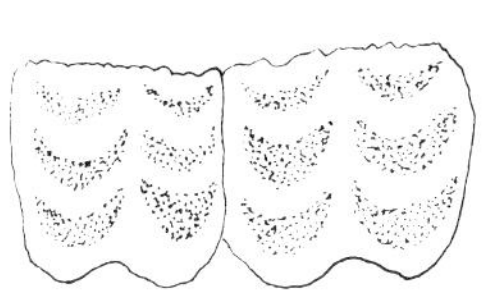

possum molars

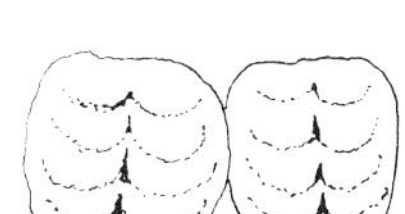

kangaroo molars

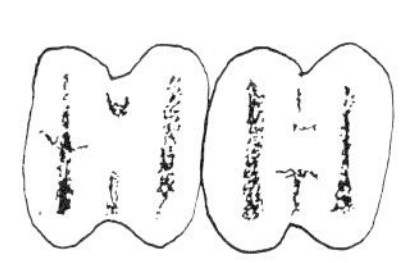

pig molars

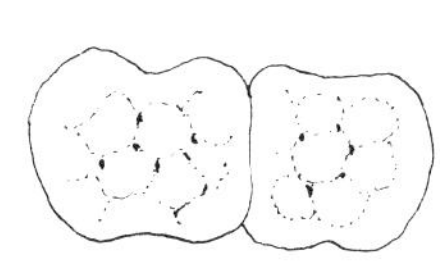

Some native rodents' skulls are difficult to tell from introduced species. House mice are relatively easy to pick because their upper incisors have a distinct notch on the back which native pseudo-mice lack , (See picture next page).

Another way of telling the difference between certain rodents is to look at the angle of the upper incisors. Some such as House Mouse and Water-rat are vertical. Some slope backwards (Pseudomice, Common Rock-rat and Black Rat). Others, such as Forrest's Mouse, slope forward.

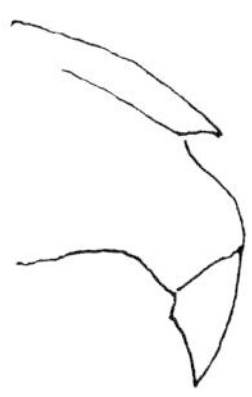

House Mouse incisors

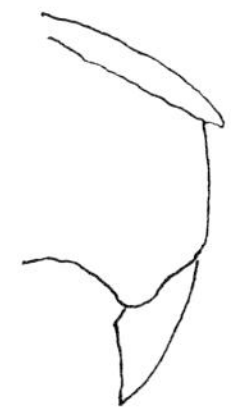

Common Rock-rat incisors

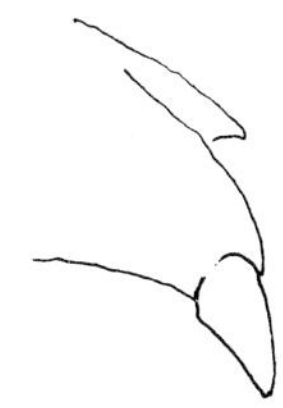

Forrests's Mouse incisors

Water-rat has two molars on either side of both jaws. Other rodents generally have three molars.

CARNIVOROUS MAMMALS

The teeth of cats and dogs may look similar to those of the larger carnivorous marsupials. However they have only three pairs of upper incisors, whereas most carnivorous marsupials have four or five. Bandicoots are an exception - they have three pairs, like cats and dogs.

Bandicoots and other carnivorous marsupials have four pairs of molars. Dogs have two upper and three pairs of lower molars, and cats have three upper and two lower. Bandicoots' molars are relatively flat for grinding plant material. Cats and dogs have blade-like molars. The molars of small carnivorous marsupials are pointy, indicative of their largely insectivorous diet.

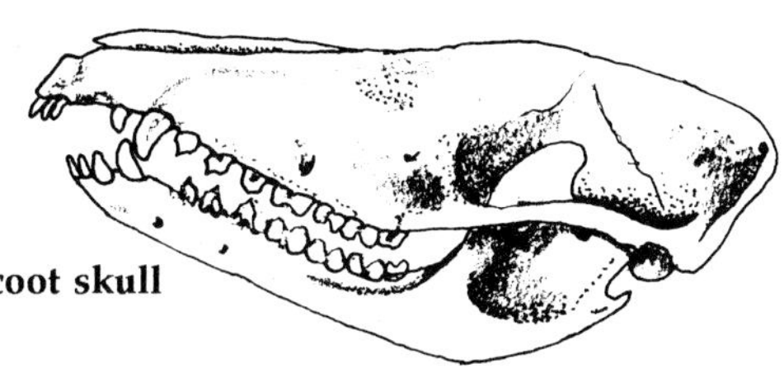

bandicoot skull

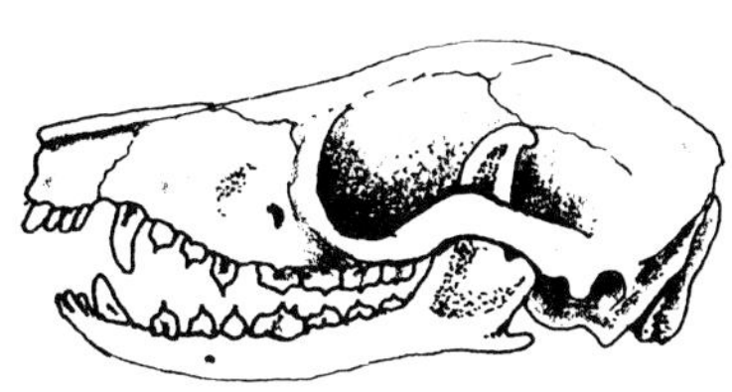

quoll skull

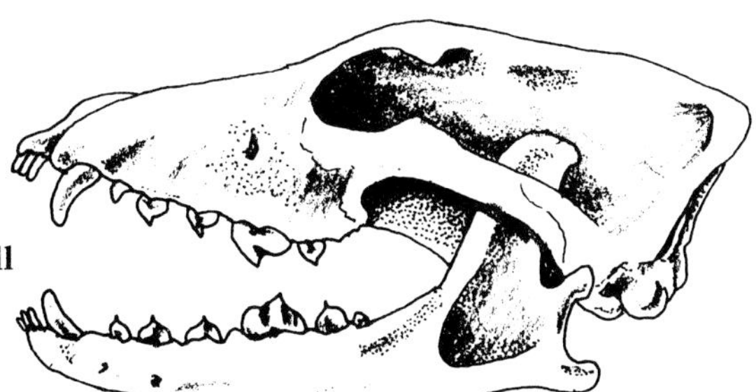

dog skull

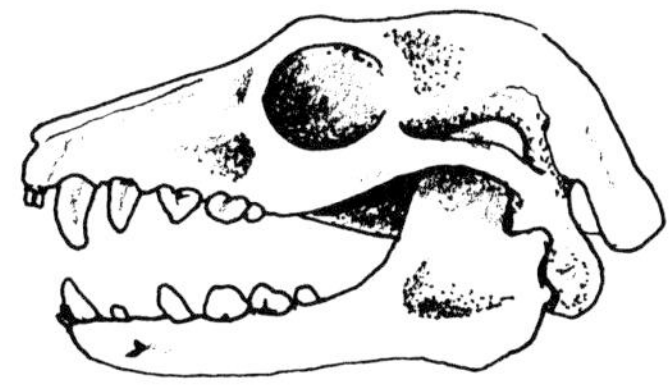

flying-fox skull

Bats

Bats have been included here because most are insectivorous and all, even nectar and fruit feeders, have canine teeth. Therefore their skulls could be confused with those of carnivores. Bats usually have two or less pairs of upper incisors. Flying-foxes have two pairs of upper and lower incisors. Ghost Bats have no upper incisors and two pairs of lower. Microchiropterans vary, even in the same genus. For instance some taxa may have one or two pairs of upper incisors and two or three pairs of lower ones. However bats always have less teeth than the typical mammal.

Skulls

Mammal skulls are basically similar (apart from echidnas and whales and a few others), although skulls of marsupials tend to be much flatter than those of other mammals. Compare the bandicoot skull with the dog skull.

However some of the smaller structures are highly variable. For instance the bone below the eye socket (zygomatic arch) differs in size between various animals, being wide in possums, kangaroos, cats and dogs, narrow in bandicoots and variable in dasyurids, rodents and bats. Mammals that eat big tough prey or fibrous vegetable matter tend to have wide zygomatic arches. The zygomatic arch of those that eat small, or relatively tender prey is rather thin, or may even be missing, as in several of the small insectivorous bats. Flying-foxes are a little anomolous, having a wide zygomatic arch even though they feed only on fruit pulp, pollen and leaves.

The coronoid process, a protuberance found on the bone behind the back teeth in the lower jaw, is also variable. It may be pointy or blade-like or it may hardly noticeable. The coronoid process provides leverage for the jaw muscles so its size is usually related to diet. Therefore it is large in carnivores and usually small in herbivores, although there are exceptions. For instance it is large in seed-eaters such Delicate Mouse, Forrest's Mouse, and several *Rattus* species.

Other bones

Echidnas have a scapula (shoulder-blade) which is quite different from that of other mammals. In fact, as you will note in the diagrams below, it is more similar to that of a lizard. Echidnas and marsupials have epipubic bones that are joined to the pelvis (picture below). These bones probably support the pouch muscles and are not found in other mammals (although some extinct groups possessed them as well). These bones are often slender and may be missing from an old skeleton.

bird scapula

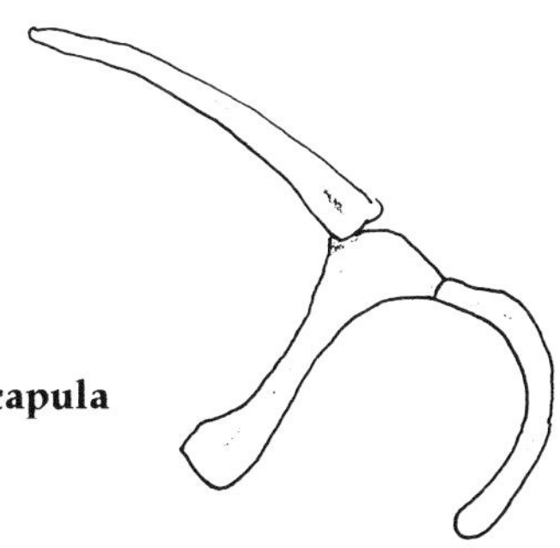

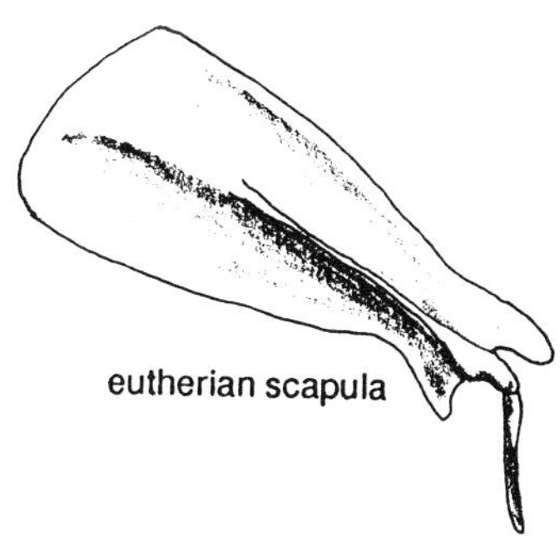

eutherian scapula

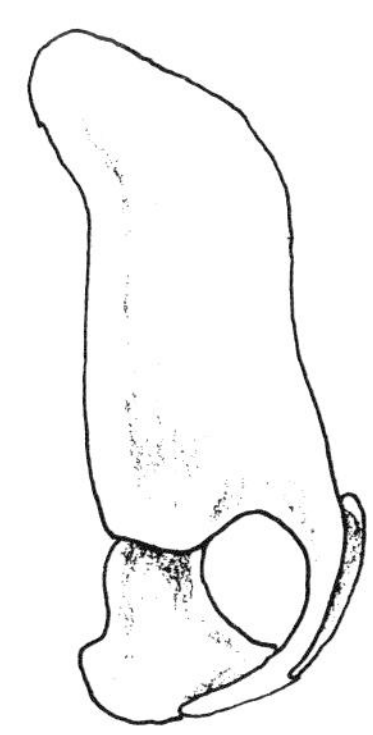

monotreme scapula

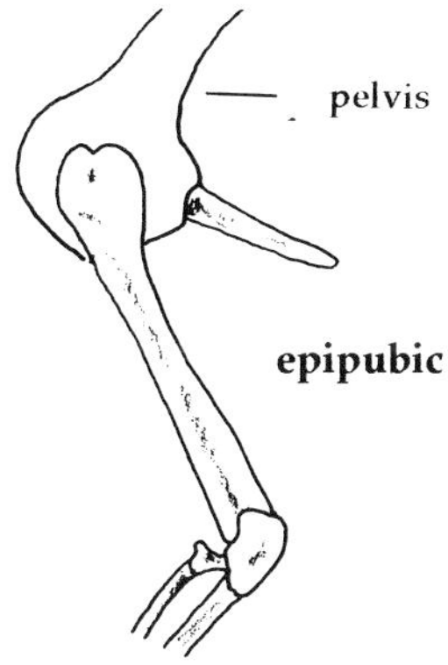

epipubic bone

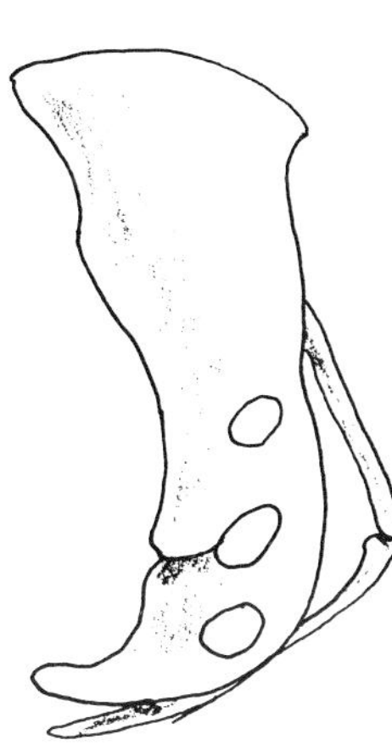

reptile scapula

Bat skeletons can be easily differentiated from those of birds, even without the skull. The most uncomplicated feature is probably the sternum. Birds (apart from emus and cassowaries) usually have a huge keeled sternum compared with bats and indeed with other animals.

Out of interest male dogs, cats, rodents and bats have a bone in their penis, the baculum. However like other small bones these usually are missing from the weathered skeleton. All mammals have epiphyses, small discs of bone, between their vertebrae.

REPTILES

Teeth

Some lizards have teeth that are fused to the jawbone (acrodont teeth). Other lizards (and frogs) have teeth which only touch the bone (pleurodont teeth). Some lizards such as agamids, have a combination of these two types. Snakes and crocodiles have a type of tooth which fits into a bony socket (thecodont teeth). Mammals have thecodont teeth as well. Turtles are toothless.

Skulls

Some reptiles, such as lizards, have many more bones in their skull than mammals. (Compare the lizard skull below with the mammal skulls, page 91 and 92).

Lizards, snakes and birds have kinetic skulls, that is, they have several joints in their skulls. Lizard skulls are usually made up of four moveable parts all joined together with cartilage Some, such as Burton's Snake-lizard, *Lialis burtoni* (see picture below) have more. Snakes may have eight points of articulation including one at the front of the lower jaw. This enables one half of the jaw to grip prey while the other moves forward to re-engage at a a further point. This is a very effective way of handling big and slippery prey (would be a great boon for human eaters of pasta and noodles). However, having so many joints means that snake skulls fall apart when the flesh rots. Reptiles, apart from crocodiles, and skinks for some reason, lack a secondary palate.

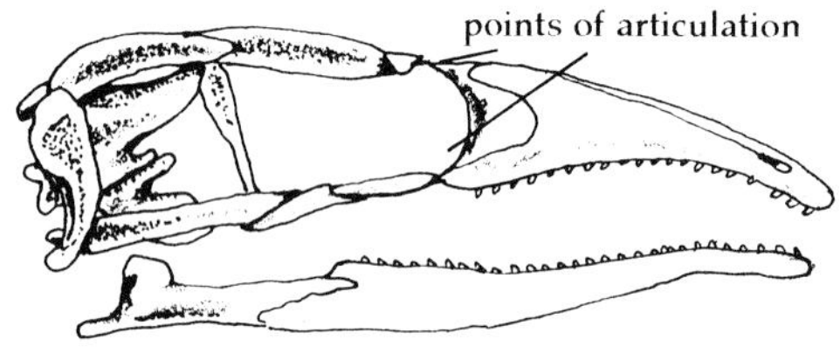

Skull of Burton's Snake-lizard

dragon lizard skull

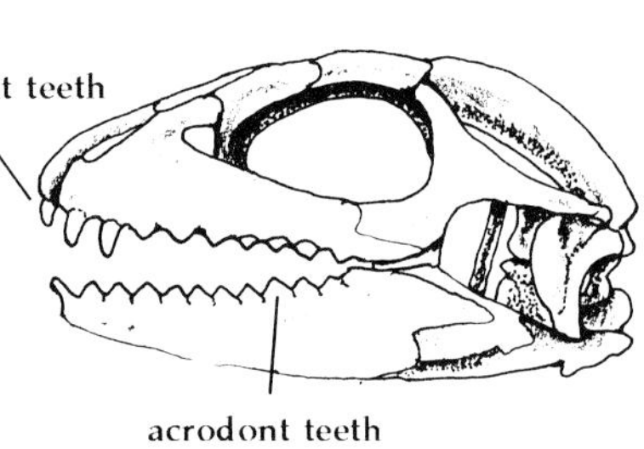

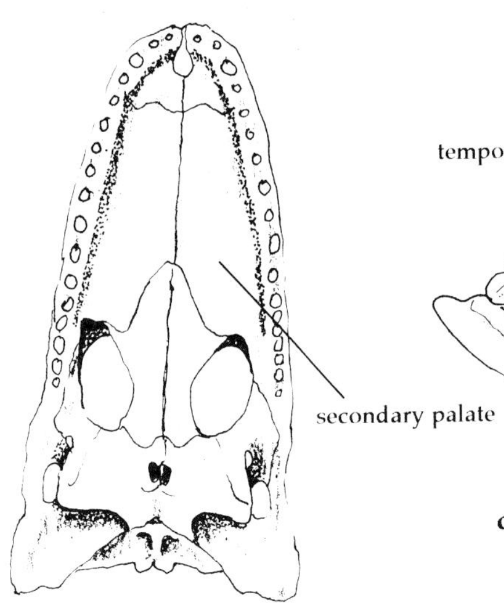

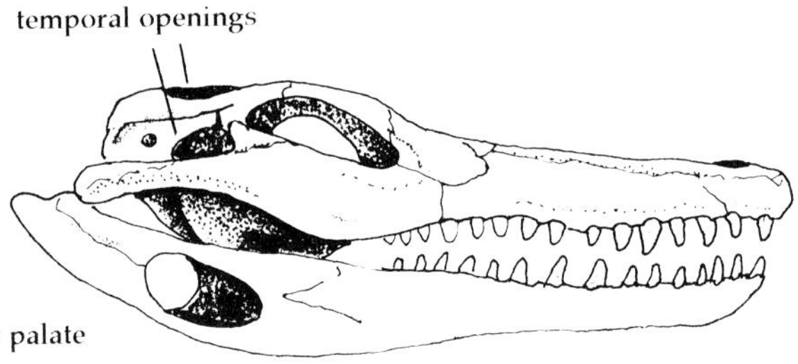

crocodile skull

Crocodiles have an unmistakeable skull with large teeth and a secondary palate. They also have two large openings on the top and sides of the skull. Estaurine crocodiles' teeth are much more differentiated than Johnston's Freshwater Crocodile. Turtle heads are rounded with no snout and there are no openings in the skull behind the eye sockets.

Other bones

Reptiles have ribs which protrude from the cervical vertebrae. Crocodiles (and skinks) have osteoderms - round bony plates under the scutes which are keratinous plates (keratin is the stuff that makes up your fingernails) that cover the dermis of crocodiles and turtles. A crocodile may have around 100 osteoderms ranging in size from a 20¢ piece to saucer-size.

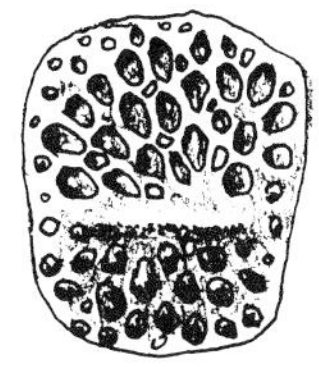

osteoderm

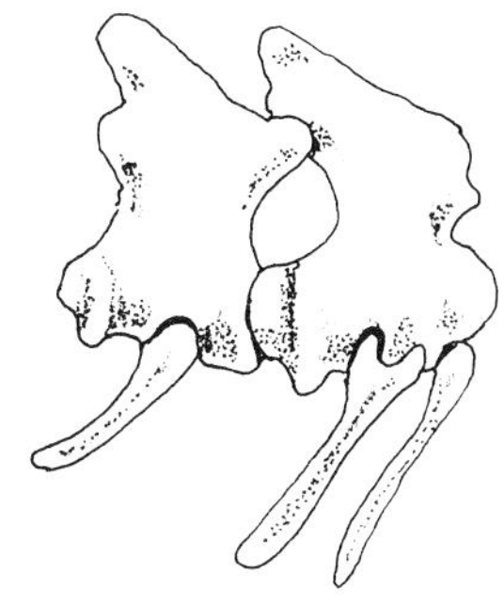

cervical ribs

Crocodiles carry pebbles in their stomachs that enable them to grind food and to act as stabilizers. These should be present in any crocodile skeleton.

Turtles are easily identifiable from their bony shells. The shoulderblade, or scapula is joined to the top part (carapace) as are the spine and the ribs. The clavicle (collarbone) is joined to the belly section (plastron). Sea-turtles' flippers lack ankle joints unlike most freshwater turtles.

Snakes lack a sternum, and all reptiles lack epiphyses, the bony discs that mammals have between their vertebrae.

TRACKS

Tracks differ according to gait. Those shown below are from walking animals.

Echidnas show long drag tracks from the long claws on their hindfeet.

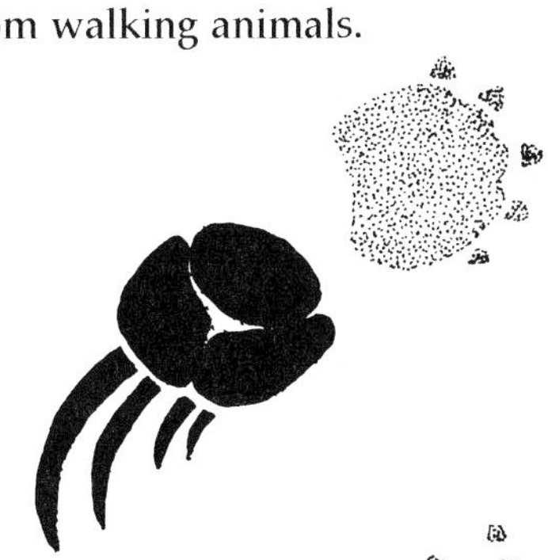

Echidna tracks

The tracks of carnivorous marsupials are all similar (except for the bandicoot), showing five digits on the forefoot and usually four toes on the hindfeet. They place their feet at an angle unlike rodents.

The tracks of rats and mice show four digits on the forefoot (the first digit is often reduced in size) and five toes on the hindfoot. Water-rat has partially webbed hind feet.

Tracks of Brush-tailed Phascogale

rat tracks

Bandicoot tracks - only three toes on the forefoot show (the other two are rudimentary). The back feet show one large middle toe and another on either side. The inner toe shows only faintly. The front feet move alternatively and the back feet move together. Kangaroos tracks are similar except that the feet move in pairs, and all five digits on the forefeet show up.

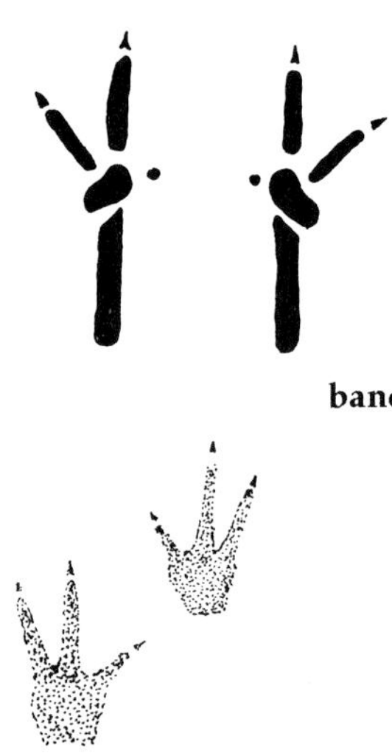
bandicoot tracks

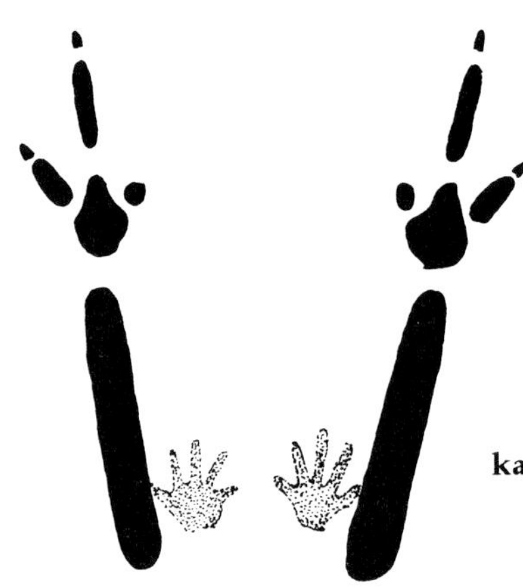
kangaroo tracks

Lizard tracks show a long and broken tail track between alternate footprints. Both forefeet and hindfeet have five digits. Those on the fore-feet are all similar but the back feet may have a short big toe and a very long fourth toe.

Crocodile feet have fivedigits on the forefeet, and four on the back feet (the fifth digit is reduced to a single bone). The front digits are free, but the hind digits are webbed. The tail leaves a very wide unbroken undulating track .

Sea turtles leave a very wide drag mark. Freshwater turtles have five digits on both front and back feet.

ANIMAL POO (SCATS)

Australian animals may be hard to see because many are nocturnal. However they do leave behind little messages in the form of poo. Drawings are roughly life- size.

Echidna scats contain the external skeletons of ants or termites (break them open and examine them with a hand lens). Northern Quoll scats are smooth or rough-textured and they stink . They may contain exoskeletons of insects, and hair. Brush-tailed Phascogale poo is also rough-textured and may contain hair. They smell like sweaty socks that have spent a week at the bottom of the dirty clothes basket. Bandicoot scats look like a dwarf version of human poo in colour, shape and texture with occasional blades of grass.

Sugar Glider and Northern Brushtail Possum poo is found at the base of large trees. Agile Wallaby and Antilopine Wallaroo poo (not depicted) is similar to that of Short-eared Rock-wallaby. However both are larger and lack a pointy end. All kangaroo, and most possum and rodent poo, consists of vegetable matter, and does not usually have an unpleasant smell. Water-rat scats are the exception. They are really on the nose, because of this animal's carnivorous diet.

Dugong scats can sometimes be found along the shore at Casuarina Coastal Reserve. They are cyclindrical, dark-brown or olive-brown in colour, and consist of plant fibre.

Crocodiles leave a white mess on the ground, much like a big waterbird's dropping.

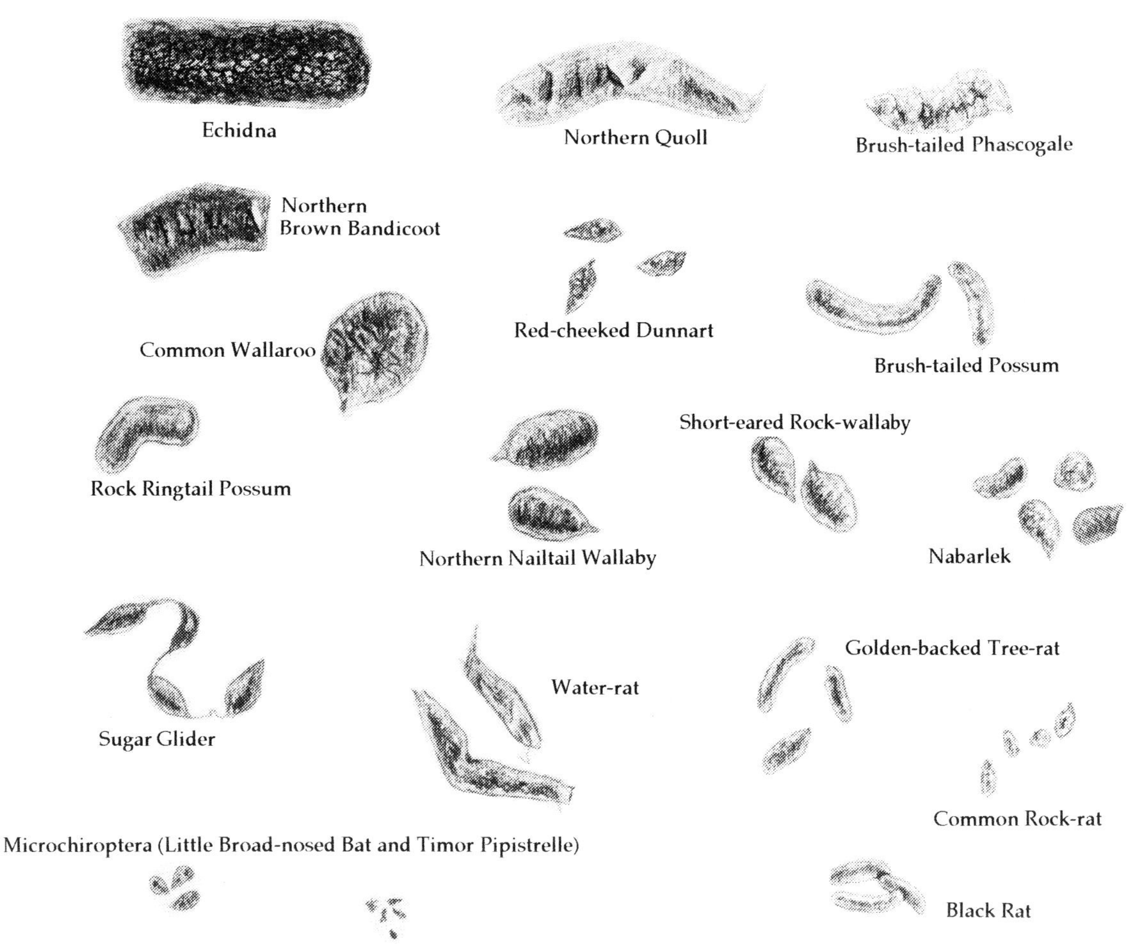

BIBLIOGRAPHY

This list of books and papers is for those who want more information..

Books by local people
Polishuk, N & D. Lockwood (1987), *Life on the Daly River*, pub. by Nancy Polishuk

Spillett, PG (1979). *Forsaken Settlement*. Lansdowne Press.

General
Dixon, JM & L Huxley (1985) *Donald Thomson's Mammals and Fishes of Northern Australia*, Nelson, Victoria.

M. Kennedy, *Australia's Endangered Species* - Simon & Schuster, Australia.

Palaeontology
Archer, M; Hand, S and H. Godthelp (1991), *Riversleigh, The Story of Animals in Ancient Rainforests of Inland Australia*, Reed Books, NSW.

Flannery, TF (1989), 'Origins of the Australo-Pacific Land Mammal Fauna', *Australian Zoological Reviews*, (1).

Murray, PF (1987) 'Plesiosaurs from Albian aged Bathurst Island formation siltstones near Darwin, Northern Territory, Australia'. *The Beagle, Records of the Northern Territory Museum of Arts and Sciences*, 4 (1): 95 - 102.

White, M.E. (1990), *The Nature of Hidden Worlds*, Reed Books, NSW.

Mammals
Bouliere, F. (1964). *The Natural History of Mammals*, 3rd edn. Knopf. New York.

Bradley, AJ; Kemper, CM; Kitchener, DJ; Humphrey, WF; & RA How,(1987), 'Small Mammals of the Mitchell Plateau Region, Kimberley, Western Australia', *Australian Wildlife Research*, 14: 397-413.

Flannery, T. (1990) *The Mammals of New Guinea*, Robert Brown & Assoc., Australia.

Maxson, LR; Sarich, VM & AC Wilson (1975) 'Continental drift and the use of albumin as an evolutionary clock'. *Nature*, 255.

Oftedal, O.T. (1984) Milk composition, milk yield and energy output at peak lactation: a comparative review. *Symposia of the Zoological Society of London*. 51 33-85.

Strahan, R. (1991) The Complete Book of Australian Mammals, Angus & Robertson, Aust.

Monotremes
Calaby, JH & K Keith (1974) 'Mammals' (in) *Fauna survey of the Port Essington district, Cobourg Peninsula, Northern Territory of Australia*, Ed. HJ Frith and JH Calaby. CSIRO Div. *Wildlife Research Technical paper* No. 28

Cribb, J. (1993); 'Platypus spurs on venom research', *The Weekend Australian*. June, 23-24.

George, GG (1973) 'Land mammal fauna', *Aust Natural History*, 12.

Griffith, M. (1968) *Echidnas*, Pergamon Press.

Griffith, M. (1978) *The Biology of Monotremes*, Academic Press, New York.

Haacke, W. (1885). 'On the marsupial ovum, the mammary pouch, and the male milk glands in *Echidna hystrix'*. *Proceedings of the Royal Society of London*, Series B. 38, 72-4

Murtagh, C.E. & G.B. Sharman, (1977) 'Chromosomal sex determination in monotremes.' In *Reproduction and Evolution*, ed.J.H. Calaby & C.H. Tyndale-Biscoe, pp. 61-6. Canberra: Australian Academy of Science.

Rismiller, P. (1991) 'The Echidna', *Scientific American*, Feb. Vol. 264, No. 2.
Any other papers by Dr. Rismiller.

Vogel, P. (1981) 'Occurrence and interpretation of delayed implantation in insectivores', *Journal of Reproduction and Fertility*, Supplement 29, 51-60.

Marsupials

Archer, M (1973), 'New information about the Quaternary distribution of the thylacine (Marsupialia, Thylacinidae) in Australia'. *Journal of the Royal Society of Western Australia,* Vol. 57, Part 2, August, 1974.

Archer, M (ed), (1982), *Carnivorous Marsupials.* Royal Zoological Society of NSW, Sydney.

Begg, R.J. (1981), 'The Small Mammals of Little Nourlangie Rock, NT. Ecology of *Antechinus bilarni,* the Sandstone Antechinus (Marsupialia: Dasyuridae)', *Australian Wildlife Research,* (8).

Begg, R.J. (1981) 'The Small Mammals of Little Nourlangie Rock, NT. Ecology of *Dasyrus hallucatus,* the Northern Quoll (Marsupialia: Dasyuridae)', *Australian Wildlife Research,* (8) 73-85.

Hunsaker, D. & D. Shupe (1971), Behaviour of New World marsupials' (in) *Biology of Marsupials,* ed. D. Hunsaker II Academic Press, New York, pp. 279-347.

Ingleby, S (1991), 'Distribution and Status of the Northern Nailtail Wallaby, *Onychogalea unguifera* (Gould, 1841), *Wildlife Research,* 1991, 18 655-18

Lyne, AG (1974), 'Gestation Period and Birth in the Marsupial *Isoodon macrourus', Australian Journal of Zoology,* 1974, 22, 303-9.

Palmer, AB. A film Documentary Concept on the Tasmanian Tiger - Survival on the northern Australian mainland.

Rodger, JC and RL Hughes (1973), 'Studies of the accessory glands of male marsupials', *Australian Journal of Zoology,* 21, 303-20.

Rounsevell, DE and SJ Smith (1982), 'Recent alleged sightings of the Thylacine (Marsupialia, Thylacinidae) in Tasmania' (in) *Carnivorous Marsupials,* ed. M. Archer Royal Zological Society, NSW. Sydney.

Stodart, E. (1966), 'Management and behaviour of breeding groups of the marsupial *Perameles nasuta* Geoffroy in captivity'. *Australian Journal of Zoology,* 14. 611-23.

Tyndale-Biscoe, H. & M. Renfree (1987) *Reproductive Physiology of Marsupials.* Cambridge University Press, Great Britain.

Tyndale-Biscoe, CH & PA Janssens (Eds), (1988). *The Developing Marsupial,* Springer-Verlag, Germany.

Bats

Churchill, S.K. (1991) 'Distribution, Abundance and Roost Selection of the Orange Horsehshoe-bat, *Rhinonycteris aurantius,* a Tropical Cave-dweller'. *Wildlife Research,* Vol. 18 No. 3.

Hill, JE and JD Smith (1985) *Bats, A Natural History,* British Museum, London.

Kitchener, DJ (1973), 'Reproduction in the Common Sheath-tailed Bat, *Taphozous georgianus* (Thomas) (Microchiroptera: Emballonuridae), in Western Australia', *Australian Journal of Zoology,* 21, 375-89.

Molnar, R.E., Hall, L.S. and Mahoney, J.H. (1984), 'New Fossil Localities for *Macroderma',* Miller 1906 (in) New South Wales and its past and present distribution in Australia'. *Aust. Mammal Vol.* 7: 63-73.

Tidemann, CR; Priddel, DM; Nelson, JE & JD Pettigrew (1985), 'Foraging Behaviour of the Australian Ghost Bat, *Macroderma gigas* (Microchiroptera: Megadermatidae)'. *Aust. Journal of Zoology,* 33, 705-13.

Thompson, B.G. (1991), *A Field Guide to Bats of the Northern Territory,* Northern Territory Conservation Commission. Gov. Printer N.T.

Rodents

Barnett, SA (1976), *The Rat: a Study in Behavour.* Aust. National University Press, Canberra.

Barrow, GJ (1964), '*Hydromys chrysogaster* - some observations'. *Qld. Naturalist.* 17.

Baverstock, PR; Watts, CHS; Gelder, M. &M. Adams (1978), 'A genetical approach to the taxonomy of Australian *Melomys* (Muridae), *Aust. Mammal Soc. (Inc.) Bulletin,* 5 (1).

Baverstock, PR; Watts, CHS; & JT Hogarth (1977), 'Chromosome evolution in Australian rodents 1. The Pseudomyinae, the Hydromyinae and the *Uromys/Melomys* group. *Chromosoma* (Berl.) 61.

Bradley, AJ; Kemper, CM; Kitchener, DJ; Humphrey, WF; How, RA and LH Schmidtt, (1988), 'Population Ecology and Physiology of the Common Rock Rat, *Zyzomys Argurus* (Rodentia: Muridae) in Tropical Northwestern Australia' *Journal of Mammalogy,* 69 (4): 749-764

Covacevich, J and M Archer (1974), 'The distribution of the cane toad, *Bufo marinus,* in Australia and its effects on indigenous vertebrates', *Mem. Qld. Museum.* 17.

Crichton, EG (1969), 'Reproduction in the pseudomyine rodent *Mesembriomys gouldii* (Gray) (Muridae) *Aust. J. Zoology* 17.

Crichton, EG (1974), 'Aspects of Reproduction in the Genus *Notomys* (Muridae), *Australian Journal of Zoology*, 22, 439-47.

Dennis, E & JI Menzies, (1978), 'Systematics and chromosomes of New Guinea Rattus' *Aust Journal Zoology*, 26.
Dwyer, PD (1975) 'Observations on the breeding biology of some New Guinea murid rodents' *Aust. Wildlife Research* 2.

Happold, M (1976), 'Reproductive biology and developments in the conilurine rodents (Muridae) of Australia'. *Aust. Journal of Zoology* 24

Knox, E. (1976), 'Upper molar alveolar pattern of some Muridae in Queensland and Papua New Guinea', *Mem. Qld. Museum* 17.

Knox, E. (1978), 'A note on the identification of *Melomys* species (Rodentia: Muridae) in Australia'. *Journal Zoology London*. 185.

Lidicker, WZ (1973) 'A phenetic analysis of some New Guinea rodents.' *Systemic Zoology* 22.

Taylor, JM & BE Horner (1971), 'Reproduction in the Australian tree-rat *Conilurus penicillatus* (Rodentia: Muridae)', *CSIRO Wildlife Research*. 16.

Taylor, JM and BE Horner (1972), 'Observations on the reproductive biology of *Pseudomys* (Rodentia: Muridae)', *Journal of Mammalogy* 53.

Taylor, JM and BE Horner (1973) 'Reproductive characteristics of wild native Australian *Rattus* (Rodentia: Muridae), *Australian Journal of Zoology*, 21, 437-75.

Watts, CHS & HJ Aslin (1981), *The Rodents of Australia*, Angus &Robertson Publ. Australia.

Dingo
Corbett, L. (1975), 'Dingo society and its maintenance: a preliminary analysis' (in) Fox, M. (ed) *The Wild Canids, Their Systematics, Behavioural Ecology and Evolution*. Van Nostrand Reinhold, New York.

Dugong
Anderson, PK and A Birtles (1978), 'Behaviour and Ecology of the Dugong, *Dugong dugon* (Sirenia): Observations in Shoalwater and Cleveland Bays, Queensland' (in) *Australian Wildlife Research*. 5 1-23.

Reptiles- General
Cogger, HG (1992), *Reptiles and Amphibians of Australia*, Reed Publ., NSW.

Shea, G; Weigel, J; Harwood, A; Floriani, H & C Hemsley (1988), 'Notes on the herpetofauna of Mitchell Plateau, Western Australia. Results of the 1987 Australian Herpetological Society field trip to the Kimberleys', *Herpetofauna* 18 (1).

Wilson, S.K. & D.G. Knowles (1988), *Australia's Reptiles* - Collins, Australia.

Plesiosaurs and Icythyosaurs
Murray, PF (1987), 'Plesiosaurs from Albian aged Bathurst Island Formation Siltstones near Darwin, Northern Territory, Australia', *The Beagle*, Records of the Northern Territory Museum of Arts and Sciences, 4 (1) 95-102.

Crocodiles
Cogger, HG (1992), *Reptiles and Amphibians of Australia*, Reed Publ., NSW.

Grenard, S (1991), *Handbook of Alligators and Crocodiles*, Krieger, Florida.

Magnusson, W.E. and J.A Taylor (1981) 'Growth of Juvenile *Crocodylus porosus* as effected by season of hatching', *Journal of Herpetology*, 15 (2).

Magnusson, W.E. (1982), 'Mortality of Eggs of the Crocodile *Crocodylus porosus* in Northern Australia', *Journal of Herpetology*, Vol. 16 (2).

Webb, JW; Manolis, SC & PJ Whitehead (1987), *Wildlife Management: Crocodiles and Alligators*, Surrey Beatty & Sons, NSW, in ass. with the Conservation Commission of the Northern Territory.

Turtles
Cogger, HG (1992), *Reptiles and Amphibians of Australia*, Reed Publ., NSW.

Georges, A and R. Kennett (1989), 'Dry-season Distribution and Ecology of *Carettochelys insculpta* (Chelonia: Carettochelydidae) in Kakadu National Park, Northern Australia' (in) *Australian Wildlife Research*, 16 (3) 323-334.

King, M and P Horner (1987), 'An additional freshwater turtle species and the confirmation of *Chelodina novaeguineae* in the Northern Territory, Australia', *The Beagle*, Records of the Northern Territory Museum of Arts and Sciences, 4 (1): 83-84

Lizards
Greer, AE (1989) *The Biology and Evolution of Australian Lizards*, Surrey Beatty & Sons, NSW.

Patchell, FC and R Shine (1986), 'Feeding Mechanisms in Pygopodid Lizards: How can *Lialis* Swallow Such Large Prey?', *Journal of Herpetology*, Vol. 20 (1).

Shea, GM and GL Reddacliff (1986), 'Ossification in the Hemipenes of Varanids', *Journal of Herpetology*, Vol, 20 (4).

Snakes
Government Publication, (1986), *Atlas of Elapid Snakes of Australia*, Australian Flora and Fauna series, No. 7.

Shine, R. (1980), Ecology of Eastern Australian Whipsnakes of the Genus *Demansia*. *Journal of Herpetology*, 14 (4)

Shine, R. (1981),' Ecology of Australian Elapid Snakes of the Genera *Furina* and *Glyphodon*', *Journal of Herpetology*, 15 (2).

Shine, R and J. Covacevich (1983), 'Ecology of Highly Venomous Snakes: the Australian Genus *Oxyuranus* (Elapidae'), *Journal of Herpetology*, Vol 17 (1).

Shine, R. (1993), *Australian Snakes, A Natural History*, Reed Books, NSW.

Frogs
Cogger, HG (1992), *Reptiles and Amphibians of Australia*, Reed Publ., NSW.

Tyler, M. J. (1984), *There's a Frog in my Throat (deleted) Stomach*, William Collins, Sydney.

Tyler, M.J. & M. Davies (1986), *Frogs of the Northern Territory*, Univ.of Adelaide, for N.T. Conservation Commission.

Introduced animals.
Tulloch, DG (1979), 'The Water Buffalo, *Bubalus bubalis*, in Australia: Reproductive and Parent-offspring Behaviour, *Australian Wildlife Research*, 1979, 6 265-87.

Acknowledgements

I wish to thank my partner Brian Whitehead for support which went beyond the call of duty. He did not bat an eyelid, even when he found me and Rowan dissecting a three metre python on the coffee table. Few would agree good humouredly, to boil a rotten bandicoot on the barbecue so that I could retrieve the bones. This book owes a lot to my gorgeous son, Rowan, who urged me to dissect the snake on the table so he could 'see how it works'. His interest has been a driving force.

My sincere thanks also go to Ita Buttrose and Penni Tastula for liking the 'things (I) do', even if champagne suppers at the sewerage ponds and snake-sexing may not be their first choice of leisurely pastimes. I'll never forget Jan Hartwig for the encouragement she has given nor Sue Fraser-Adams who keeps telling me I'm wonderful. This book would have remained unprinted if it had not been for Gary Coleman and his staff, and Michael Bollen of Wakefield Press.

My adopted Aboriginal son, Peterson and his wife, Stephanie, gave me the Gunwinggu information in the book, going back to their elders to check details. Thank you for sharing your knowledge, and for not laughing at my pronounciation as you have tried to teach me your language over the years.

Dr. Peggy Rismuller and Mike of Pelican Lagoon, Kangaroo Island were great. It was wonderful to find other people in the world who sex animals for visitors. However I never dreamed when I asked for a picture of an Echidna's penis that I would get an autographed A4-sized, full technicolour poster!

Thanks also to Barbara Evans of the Northern Territory Tourist Commission, Brian Measey and

Mike Nixon of Territory Rent-a-Car, and Grace and Gus of Park Lodge for their support over the years when the rest of the tourism industry seemed to regard me as either a troublemake, or quite mad!

Thanks to Lee Moyse of Berry Springs Wildlife Park who supplied me with samples of poo, and to the keepers who collected it. I am grateful to Paul Horner of the Northern Territory Museum who went to some trouble to read the draft for me and to provide further information. Also to Hilary Thompson who also read and commented on this book, and to Dr. David Wigston with whom I first discussed the book and who encouraged my interest in science.

Col Myers of Mt Isa Mines was very supportive and gave me permission to use records from environmental surveys carried out at McArthur River. Likewise Newcrest gave me permission to use Coronation Hill records, and Terry Strapp of Zapopan, the Mt Todd records. Thank you to Owen Price and John Woinarski for verbal permission to use their Conservation Commission records.

Thank you also to Keith Martin, my colleague on several bush trips who shared his avid interest in reptiles and football (the latter I could have done without) and, as I remember, only once got cross with me, when I discovered that his precious Bundaberg rum made wonderful animal bait. And to all those friends, both locals and visitors, scientists and amateurs, who have put up with my research. Thank you all for your interest, patience, earthy humour and searching questions and comments over the years.

PHONE NUMBERS

Australian Conservation Agency (formerly Australian National Parks and Wildlife)	Darwin	815299
	Kakadu	799101
Consumer Affairs		895183
Darwin City Council		822511
Northern Territory Conservation Commission	Darwin	895511
	Katherine	73 8770
Northern Territory Tourist Commission		893900
Minister for Conservation		896501
Minister for Tourism		896277

ECHIDNA WATCH

Ancestors of the ECHIDNA roamed the planet along with the dinosaurs.
Dinosaurs died out but the ECHIDNA still survives today!
What are their secrets? What are their numbers?
Can they continue to cope in a fast changing environment?

This amazing egg-laying mammal lives only in Australia and New Guinea. Two hundred years after it was first noticed by European scientists we still know very little about its natural history.

Echidnas were once wide spread across Australia, but no one knows how many are still around today.

They are shy, secretive creatures and science needs your help to find out where and how many echidnas are currently being seen. Are they becoming threatened?

After evolving for 110 million years it is time to explore and conserve the secrets of their survival. It would be a tragedy to lose yet another of Australia's unique animals.

Next time you go into the bush keep an eye open for ECHIDNAS - fill in the survey from and become part of ECHIDNA WATCH.

YOUR OBSERVATIONS COUNT. PLEASE HELP!!

FOR FURTHER INFORMATION CONTACT:

ECHIDNA CARE INC.
8 King Edward Ave.
HAWTHORN SA 5062

echidna care

supporting research for survival

ECHIDNA WATCH
First Nation-Wide Echidna Survey

Have you seen an echidna? YES/NO

If so, where? (include closest town and state)..................................

When? (date and time of day)..................................

How many animals?..................................

Description of animal (size, colour)..................................

What type of country?..................................

What was it doing when you saw it? ....

Do you frequently see echidnas? YES/NO

How often?..................................

Have you seen a dead echidna? YES/NO

If so, where?..................................

When?..................................

Road Kill/Other?..................................

Description of animal..................................

Type of country?..................................

Are you prepared to make regular observations in your area? YES/NO

Comments..................................

(Optional) Name, Address..................................

YOUR OBSERVATIONS COUNT!! PLEASE HELP.

(please copy and pass out to your friends)

Send completed form to:

ECHIDNA CARE INC. 8 King Edward Ave. HAWTHORN SA 5062

echidna care

supporting research for survival